Biomembranes

Biomembranes

Fundamentals in Relation to Human Biology

James S. Beck, M.D., Ph.D.

Faculty of Medicine
The University of Calgary

Hemisphere Publishing Corporation

Washington New York London

McGraw-Hill Book Company

New York St. Louis San Francisco Auckland Bogotá
Düsseldorf Johannesburg London Madrid Mexico
Montreal New Delhi Panama Paris São Paulo
Singapore Sydney Tokyo Toronto

NOTICE

Medicine is an ever-changing science. As new research and clinical experience broaden our knowledge, changes in treatment and drug therapy are required. The editors and the publisher of this work have made every effort to ensure that the drug dosage schedules herein are accurate and in accord with the standards accepted at the time of publication. Readers are advised, however, to check the product information sheet included in the package of each drug they plan to administer to be certain that changes have not been made in the recommended dose or in the contraindications for administration. This recommendation is of particular importance in regard to new or infrequently used drugs.

BIOMEMBRANES

Fundamentals in Relation to Human Biology

1 2 3 4 5 6 7 8 9 0 DODO 7 8 3 2 1 0 9

Library of Congress Cataloging in Publication Data

Beck, James S

Biomembranes.

Bibliography: p.

Includes index.

1. Membranes (Biology) I. Title.

[DNLM: 1. Biology. 2. Cell membrane.

QH601 B393b]

QH601.B35 1979 574.8'75 79-13601

ISBN 0-07-004263-2

This book was set in Press Roman by Hemisphere Publishing Corporation. The editors were Judith B. Gandy and Jeanne Bockel; the production supervisor was Rebekah McKinney; and the typesetter was Ronald L. O'Dwyer. R. R. Donnelly & Sons Company was printer and binder.

Contents

Preface

The more central a concept is to active scientific investigation, the more transient our knowledge of it. The biomembrane is a case in point. It is an identifiable, unique piece of the physical and biologic world. Much like the old idea of the gene, it is also so rich a concept that it guides investigations and formulations in virtually every field of biology. This makes it well worth several books. However, the huge amount of work and the high rate of accumulation of new information threaten to date any book on biomembranes very quickly.

It is intended that the emphasis on fundamental physical and chemical laws and well-established observations from cell biology will give the book a long life despite the rapid development of the field. The focus on fundamentals and generalization where possible–justified by the goal of lasting usefulness–lead as well to a minimization of citations and discussions of very specific reports. This should facilitate reading for the basic ideas. The loss that results from lack of citations is compensated for somewhat by the Annotated Bibliography, which provides access to the literature through a few selected books and review articles. In some cases these entries were chosen also because other articles in the volumes cited are relevant to closely related interests.

The primary purposes of this book are: (1) to introduce some of the terminology used in writing about biomembranes, (2) to help the reader establish an adaptable conceptual base for learning more about biomembranes and cell biology, and (3) to relate diverse biologic phenomena to common ideas as basic and as lasting as possible.

The secondary purposes are: (1) to serve as a supplementary text in biology courses–in particular in cell biology and physiology courses, (2) to provide researchers in related fields and practitioners in fields such as veterinary and human medicine with a quick background to a central and burgeoning scientific endeavor, and (3) to illustrate for the general reader a productive way of looking at biologic phenomena.

The reader will have heard it said that there are more scientists practicing now than have lived in all the generations past. This, of course, depends on the definition of "scientist." Nevertheless, there are a lot of us, and our insights, attitudes, and motivations vary considerably. I hope this book conveys, to some readers at least, the finer spirit of inquiry some of us feel. There is more than one sphere in which the human spirit can soar; science is certainly among them. The mind can experience beauty in a variety of ways. I hope the reader can find in reading these pages the beauty and satisfaction of understanding in one's own mind.

Any scientist-author must acknowledge her or his colleagues–from the beginning to the present–who have provided information and intellectual stimulus. It is particularly important here to make that acknowledgment explicit, since I have–for reasons given above–omitted citations for most of the findings used in this book. More specifically, I want to acknowledge the aid of Mrs. Dolores Foster, who prepared the typescript, not only as a highly skilled typist but also as an intelligent reader. Most of the original drawings were done by Will Symmes and Trudy Jorgenson. Art work, graphics, and photography were done in the Medical Instructional Resources A/V Production Unit of the Faculty of Medicine of The University of Calgary for use in the medical curriculum. The initial scaled drawing for Figure 2-17 was done by Marian Beck. Judith B. Gandy was an informed and congenial editor.

James S. Beck

Biomembranes

Chapter 1

Perspective

One of the purposes of this book is to prepare the reader to follow the development of our collective understanding of biologic phenomena. This requires, in addition to an appreciation of the basic principles, a readiness to incorporate new information, to alter ideas, and to reject some interpretations for better ones. In short, we need both a realistic view of the limitations of people as scientists and a certain degree of humility. Nothing does more for our humility and intellectual flexibility than a look at how people worthy of our respect have been moved to change their views. In this spirit, Section 1-1 is a brief look at the historical development of some ideas that have brought the biomembrane into prominence. Section 1-2 presents a resulting conceptualization of the biologic world. Then, in Section 1-3, a definition of biomembrane and a classification of forms are given.

1-1 HISTORICAL OVERVIEW

The history of biology begins with the arrival of people as observers and systemizers. For our purposes, however, we begin with the view of animals and plants as organisms. It has been known since ancient times that the larger organisms were composed of parts that clearly differed from one another—that is, organs. The delineation of the components of organisms beyond macroscopic organs and fluids had to await the development of the microscope in the seventeenth century.

Before the use of the microscope there were two main lines of thought about the physical behavior of organisms. One was the mechanical view supported by work such as that of Harvey on circulation (1628) and that by Descartes in his posthumously published work "De homine" (1662). Descartes' mechanical view of the animal organism (cords, tubes, and fluids) was widely accepted by contemporary biologists. But, at the same time, another line of thought grew from the failure of such mechanical devices to explain all that living things were seen to do. This view rested on chemistry, which was then in a very early stage of development. The chemical view of biologic processes was taken by Paracelsus in the sixteenth century. With fermentation as the prominent model, contemporaries of Descartes such as van Helmont and Sylvius tried to explain physiological processes as chemical.

In the seventeenth century biology was shaken by a technological advent—the microscope. An enormous amount of work was done within a few decades in describing the microscopic structure of organisms. Recognition of a ubiquitous unit of structure, and the assignment of the name *cell* to it, occurred at this time. But it was more than a century later—in 1805—that Oken asserted that all organisms consist of cells. Then in 1839 Schwann's "Microscopical Researches on the Similarity in Structure and Growth of Animals and Plants" appeared. Schwann held that the organism is composed entirely of cells and substances emanating from cells and that the cells live independently, but that this independence is subservient to the living state of the organism as a whole. Further, Schwann divided the capacities of cells into two groups: (1) those related to the combination of molecular components into the structure of the cell; and (2) those that derive from chemical changes. (Schwann applied the term *metabolic* to the latter category.) This conceptual dichotomy has persisted to this day and is represented in the way biology has divided into subdisciplines. The continuous restructuring of biology can be understood only in this historical context.

During the nineteenth century the behavior of cells was further described in terms of chemical transformations in light of developing techniques and understanding in chemistry. And the burgeoning study of animal organ function revealed many examples of how the behavior of organs and organisms could be explained in terms of cellular processes. Morphological investigation continued, and improved instruments made possible better visualization of the detailed structure of cells themselves. The concept of the cell as a sac bounded by a more or less flexible barrier—the cell, or plasma, membrane—and containing an aqueous solution and some particles (organelles) such as the nucleus was a generally accepted working model when Overton in 1895 reported on permeability properties of the plasma membrane of various cells. Overton noted that the permeability to substances generally increased as the lipid solubility increased and concluded that the bounding membrane is, for the most part, probably lipid. This theory was followed by others in the twentieth century and led to our current picture of the cell membrane, as discussed in Chapter 2.

In this century a virtual explosion of research and information occurred in biochemistry laboratories. Molecules of the cell contents were identified, their structures specified, their energies determined, and their physiological functions characterized. The composition of biologic systems was taken far beyond the limitations of the microscope. Furthermore, the chemical reactions relating these molecular components were described, and specific catalysts—recognized as proteins and called *enzymes*—were found to govern the rates of these reactions.

But most of this imposing bulk of information was obtained at a sacrifice. The spatial (and sometimes temporal) organization of the systems being studied was lost in the process. A chemical engineer interested in biology once remarked that it was as if an engineer, in an effort to understand what was done in a manufacturing plant, were to bulldoze it flat, mix up the pieces, and then proceed to sift through them. Nevertheless the understanding derived from biochemical studies on extracts of broken cells and organelles was penetrating and powerful. And in the fourth, fifth, and sixth decades of this century technology developed to permit examination of the spatial organization of cells.

The development of electron microscopy enabled biologists to describe the structural features of cells much more deeply than did light microscopy. Great detail in the form and composition of organelles such as the nucleus and the endoplasmic reticulum was seen. Beyond this, the membranous components of the organelles and the plasma membrane—regarded as an organelle itself—were seen to have discernible structure. These developments are discussed in Chapter 2. Simultaneously, separation techniques, particularly ultracentrifugation, were developed, which permitted a correlation of biochemical observations with morphological observations. Thus the vast amount of knowledge of biochemistry could be combined with the detailed structural information. The consequences of such a spatiotemporal view of the chemical functions of cells appear throughout this book.

During this same period developments from molecular, atomic, and nuclear physics gave us a variety of noninvasive techniques for observing molecules and atoms in biologic structures without destroying the organization of the structures. Such techniques include electron spin resonance, nuclear magnetic resonance, and certain light-scattering and light-polarizing methods.

In summary, then, over the past 300 or 400 years the concept of biologic phenomena of organisms has moved through a logical sequence of steps. In particular, animal organisms were seen, on the one hand, as machines assembled of macroscopic parts that pushed, pulled, and puffed to produce behavior and, on the other hand, as assemblages of rather large reservoirs of chemicals reacting to produce the observed behavior. The discovery of the microscopic cell then led to the picture of the organism as a variety of groups of tiny, more or less independent, living units. Later the idea was accepted that these units were foci of chemical activity that varied according to cell type. It was then recognized that the cell membrane not only defined the cell morphologically but also influenced the composition of the cell contents. With the discovery of the enzymes and their specificity, the idea of the cell as a bag of enzymes in water held sway. With more penetrating techniques (theoretical as well as experimental), the past few decades have shown the spatial and temporal organization of cells to be central to their behavior.

This elimination of the dichotomy presented by Schwann 140 years ago—the merging of chemistry and ultrastructure—is the distinctive work of current biologists at the cellular and subcellular levels. It is this history that has led to the recognition of the biomembrane, in which structure and function become inseparable, as a quintessential element in the organization and behavior of the cell.

1-2 THE BIOMEMBRANE IN THE BIOLOGIC HIERARCHY

There is much to think about in biology, and there are a variety of ways to keep the thoughts straight in one's mind. One way is to conceive of a hierarchy in which one "level" is composed of units drawn from lower levels and is itself the unit of some higher level or levels. Figure 1-1 presents such a conceptualization. As used here the terms lower and higher impute no value; they simply give direction in the hierarchy.

The meaning of level is actually quite complex but need not be elaborated further for the purpose here. Advances in understanding often amount to the replacement of arbitrary concepts like "soul" or "vital force" with conceptual bridges from level to level. The relative importance of levels in the hierarchy is a matter of taste, habit, or viewpoint. The heart

Atomic-molecular	atoms, ions, molecules
Macromolecular	large polymers: nucleic acids, proteins, polysaccharides
Supramolecular	complexes of macromolecules: ribosomes, biomembranes, etc.
Organelle	nucleus, plasma membrane, etc.
Cellular	cells
Organ	organs
Organismic	individuals of a species
Population	all or a subset of a species
Ecological	ensemble of various species and environment
Biosphere	earthly life plus environment

Figure 1-1 Hierarchical representation of animal organization. The level is named on the left and typical members of the level are given on the right. Assemblages of members of higher levels form members of lower levels usually, but not necessarily, immediately below.

of a person may be no more than a trivial means to keep cells alive, or the cell may be no more than a device to create good conditions for an enzyme to work. To each her or his own view.

Nevertheless, one can ascribe significant characteristics to the various levels. An interesting feature of the three levels macromolecular, supramolecular, and organelle is that they occupy a peculiar position in the sense of modern physics. These levels function in a space-time scale too large to manage quantum mechanically and too small for the application of continuum mechanics. And it is at the supramolecular level—centrally located in the hierarchy—that the biomembrane lies.

1-3 DEFINITION OF BIOMEMBRANE

The concept of the biologic membrane has evolved with the development and use of chemical and morphological methods. It is accepted now that a biomembrane is an orderly array of molecules of proteins, phospholipids, and sometimes cholesterol, approximately 10 nm thick. These laminae are usually curved and continuous, forming closed sacs. This definition includes all the subcellular structures biologists ordinarily consider in the context of membrane biology.

The structures included by this definition are the limiting plasma membrane of cells; most of the nuclear envelope; the endoplasmic reticulum; the major component of the Golgi apparatus; the limiting capsules of secretory granules, lysosomes, peroxisomes, and vacuoles; and the inner and outer enclosures of mitochondria (Figure 1-2). It excludes the cell wall of some bacteria and plants that resides outside the plasma membrane. It excludes the basement membrane, which is an amorphous extracellular

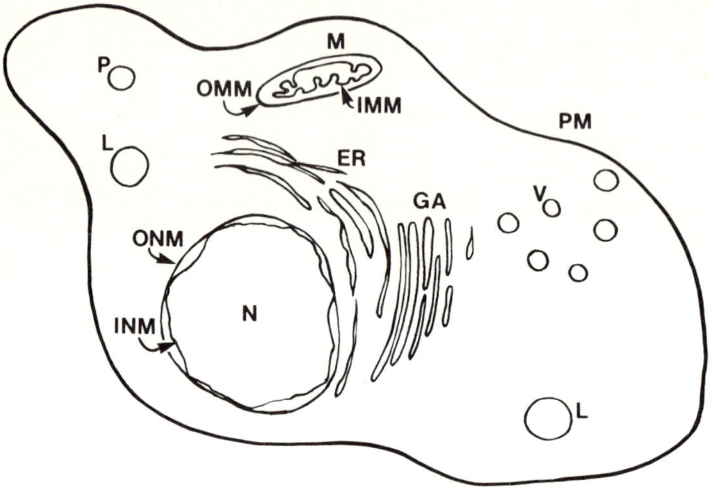

PM	plasma membrane	V	vesicles
N	nucleus	L	lysosome
ONM	outer nuclear membrane	P	peroxisome
INM	inner nuclear membrane	M	mitochondrion
ER	endoplasmic reticulum	OMM	outer mitochondrial membrane
GA	Golgi apparatus	IMM	inner mitochondrial membrane

Figure 1-2 Biomembranes of the eucaryotic animal cell.

structure composed largely of polysaccharides and related chemicals, and which is much thicker than 10 nm. The basement membrane is an important structural feature in most tissues and serves as a more or less selective barrier to body constituents. Although it is a central feature in the functional organization of some organs, it is in a class by itself, and its behavior can be analyzed more simply than that of biomembranes.

It should be mentioned that conventional usage sometimes designates as membranes some structures that do not fall within the range of the definition used here. The complex, multilayered wall of capillaries of the human vascular system is often called the "capillary membrane." This usage is quite consistent with investigations of capillary transport (much in the spirit of Chapter 3). A similar situation exists in textbook discussions of certain epithelia. Nevertheless, such structures will be regarded as systems of more than one biomembrane and are singled out for consideration in Chapter 6.

Biomembranes fall conveniently into three categories: plasma membrane, cytoplasmic membranes, and nuclear membranes. These categories can be distinguished by geometric relationships of the cell, chemical composition, ultrastructure, and physiological behavior. It is probably a mistake, however, to think of these as clear categories of independent entities. The issues—raised by improved electron microscopy—of continuity of nuclear envelope with cytoplasmic membrane, continuity among cytoplasmic membranes, and continuity between certain cytoplasmic membranes and plasma membrane are not entirely resolved. As will be seen in the discussion of biogenesis (Chapter 5), there is both morphological and chemical evidence for transition from one class to another. A conventional classification of the biomembranes of mammals is as follows:

- Plasma membrane
- Cytoplasmic membranes
  - Endoplasmic reticulum
  - Golgi apparatus
  - Mitochondrial membranes
    - Inner
    - Outer
  - Lysosome membranes
  - Peroxisome membranes
  - Vesicle membranes
    - Secretory
    - Storage
- Nuclear membranes
  - Inner
  - Outer

The plasma membrane (or cell membrane) is a closed sac forming the outermost boundary of the cell in metazoans, and lying within the cell wall in plants, some bacteria, and yeasts. It is of the order of 10 nm in thickness. The nuclear envelope of eucaryotic cells is a more or less closed sac composed of two membranes of 10 nm thickness, separated by an aqueous phase some tens of nanometers thick, and joined by special structures (nuclear pores) at widely distributed points on the sac. The endoplasmic reticulum, which is associated with synthesis of proteins and perhaps other cell constituents, is an often extensive membrane probably forming closed sacs that are flat and extended. It appears in some cells to be continuous with—or in transition from or to—both the nuclear membrane and the Golgi apparatus. The Golgi apparatus, or dictyosome, is an assembly of membranes in the form of one convoluted sac or several to many closed sacs or cisternae, usually near the nucleus, and sometimes appearing to be in transition to or from both endoplasmic reticulum and secretory granules. Lysosomes, peroxisomes, and vesicles are generally spherical and small relative to the cell (in mammals) and contain specific enzymes, chemical messengers, or stored metabolites. The membranes are closed sacs and usually 10 nm in thickness. The inner and outer mitochondrial membranes are both closed sacs, the inner membrane being contained in the outer. They are also 10 nm in thickness. The inner membrane is usually convoluted and may have originated in early eucaryotic evolution as a captive, formerly free-living entity.

These various biomembranes have been distinguished primarily by their spatial relations within the cell. There are ultrastructural differences among them, as well as differences in chemical composition. To determine the chemical composition of a membrane, the membrane must be separated from other membranes and from nonmembrane components of the cell. This problem is far from trivial. There is a logical circularity that results from the chemical component of the definition of biomembrane. Nevertheless, the collective experience with, and refinement of, methods and concepts of the biomembrane have left us with a quite satisfactory idea of what we mean by biomembrane in general. However, the chemical characterization of particular membranes requires careful control and detailed reporting of procedure.

We can demonstrate the problem with a more familiar object. Suppose we ask: "What does this particular tablecloth consist of?" The object of our query is on a table at which a lavish feast has occurred. We pull the tablecloth gently from the table and examine what we have. We conclude that this tablecloth is mostly linen but includes water, wine, bread crumbs, jelly, grease, meat, and peas. But if we had included a thorough shaking in the procedure, we would conclude that the tablecloth consists of linen, water, wine, jelly, and grease. Adding a rinse in water would lead us to conclude that the tablecloth consists of mostly linen with some grease. Adding another step, an ether rinse, would produce yet a different conclusion: 100% linen. With less directly observable objects, such as biomembranes, even more difficulty may be anticipated. However, there is more to it than composition. Our notion of the tablecloth, for example, includes ideas about its manufacture, where it is found, and the uses to which it is put.

The problem of experimental definition notwithstanding, much experience over the past few decades justifies the above definition of the biomembrane, and it is on this basis that the physical and chemical characteristics of the biomembrane are described in Chapter 2.

Chapter 2

General Properties of Biomembranes

The investigations that led to the widely accepted definition of the biomembrane given in Chapter 1 indicate that the constitutents of biomembranes are chemically of two classes: proteins and lipids. These constituents have some common physical properties that determine the molecular organization of biomembranes. The following discussion of components emphasizes the physical properties and interactions of the various types of molecules in order to establish a background for understanding the properties and structures of all biomembranes.

2-1 MEMBRANE LIPIDS

After an investigator has separated the biomembrane of interest from other cell components, the separation of membrane lipids from membrane proteins is achieved by taking advantage of one of the major defining properties of lipids. Lipids are soluble in organic solvents and relatively insoluble in water. By judicious choice of solvent, virtually complete separation into the two major categories is possible: protein in the aqueous phase or precipitated and lipid in the organic phase. The appropriate organic solvent may be pure (for example, chloroform) or a mixture (for example, chloroform-methanol, 2:1). The various methods of analytical chemistry such as elemental analysis, specific reactivity, and chromatography furnish the data from which specific structures are deduced.

The lipids found vary with type of membrane and with species. The human red blood cell (erythrocyte) membrane is well characterized and serves as a useful example. Over 99% (by mass) of the lipid is either cholesterol or one of the phospholipids. The phospholipids include sphingomyelin and several phosphoglycerides.

Most of these phospholipids are phosphoglycerides. In this class of compound two fatty acids are bonded by esterification of the carboxylic acid groups of the fatty acids with two adjacent hydroxyl groups of the glycerol. The third substitution is with a phosphoryl alcohol or its derivative. The structures of phosphoglycerides of the human erythrocyte are shown in Table 2-1. There are five different *polar head groups*: hydroxyl (phosphatidic acid), serine (serine phosphoglyceride), inositol (inositol phosphoglyceride), ethanolamine (ethanolamine phosphoglyceride), and choline (choline phosphoglyceride).[1] At normal human plasma H^+ concentration (pH 7.4), the choline phosphate group is a zwitterion with a single unit positive charge at the amino group and a single negative charge at the phosphate group; it is a net neutral, but quite polar, group. Ethanolamine phosphoglyceride can form a (+, −) zwitterion, but it is net negative at pH 7.4. Serine phosphoglyceride forms a (+, −, −) net negative ion. The polar head group of inositol phosphoglyceride is a hexahydric alcohol rather than an amine; it is net negative at pH 7.4.

The fatty acid constituents of the phosphoglycerides are variable within a membrane, within an organism, and among species. In the human erythrocyte, these are, for the most part, 18-carbon or 16-carbon fatty acids. As will be seen, the existence of double bonds in these long hydrocarbon chains is a determinant of physical properties critical to membrane behavior. In general the natural fatty acids have even numbers of carbon atoms and do not branch. Those found in the human erythrocyte membrane are listed by common name in Table 2-2 together with structures and a conventional notation to indicate structure with respect to numbers of double bonds. The phosphoglycerides of the human erythrocyte normally include one saturated (containing no double bonds) and one unsaturated (containing one or more double bonds) fatty acid group. The unsaturated fatty acid is usually esterified at the middle carbon of the glycerol core.

[1] Common names of the last four groups are, respectively, phosphatidylserine; phosphatidylinositol; phosphatidylethanolamine; and phosphatidylcholine (also called lecithin).

Table 2-1 Major Phosphoglycerides of Human Red Cell Membrane

Names	Structure	Usual fatty acid groups (R_1, R_2)
Phosphatidic acid	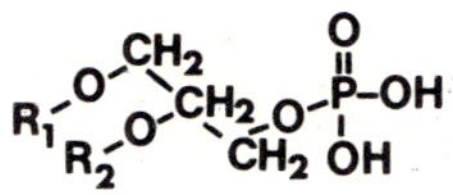	—
Serine phosphoglyceride (phosphatidylserine)	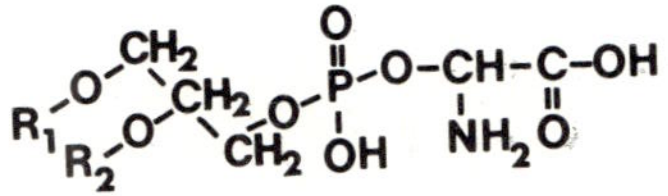	18:0, 20:4
Inositol phosphoglyceride (phosphatidylinositol)	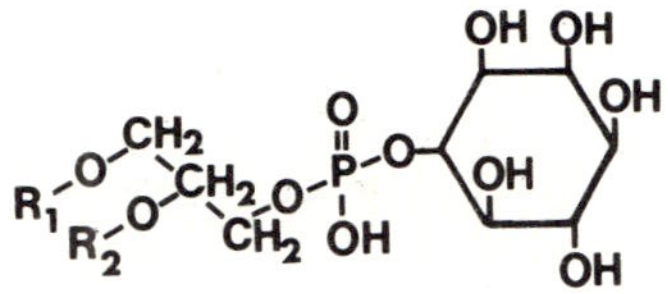	—
Ethanolamine phosphoglyceride (phosphatidylethanolamine)	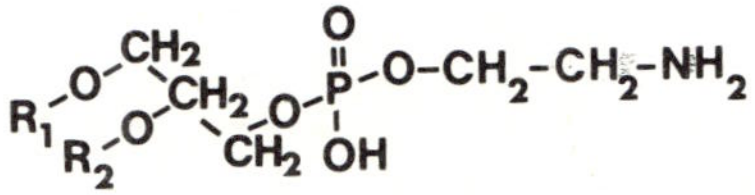	16:0, 18:0, 18:1, 20:4
Choline phosphoglyceride (phosphatidylcholine; lecithin)	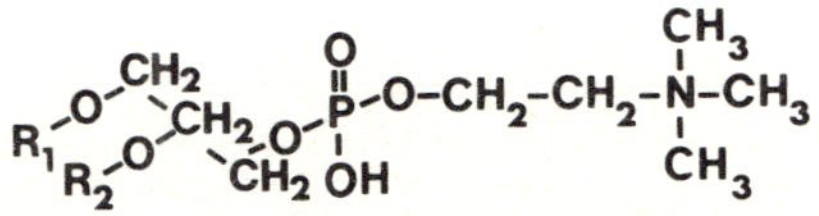	16:0, 18:1, 18:2

Table 2-2 Some Fatty Acids of Membrane Phospholipids

Common name*	Structure	Short designation
Saturated		
Lauric acid	$HOOC{-}(CH_2)_{10}{-}CH_3$	12:0
Myristic acid	$HOOC{-}(CH_2)_{12}{-}CH_3$	14:0
Palmitic acid	$HOOC{-}(CH_2)_{14}{-}CH_3$	16:0
Stearic acid	$HOOC{-}(CH_2)_{16}{-}CH_3$	18:0
Unsaturated		
Palmitoleic acid	$HOOC{-}(CH_2)_7{-}CH{=}CH{-}(CH_2)_5{-}CH_3$	16:1 or $C_{16}\,\Delta^{9}$†
Oleic acid	$HOOC{-}(CH_2)_7{-}CH{=}CH{-}(CH_2)_7{-}CH_3$	18:1 or $C_{18}\,\Delta^{9}$
Linoleic acid	$HOOC{-}(CH_2)_7{-}CH{=}CH{-}CH_2{-}CH{=}CH{-}(CH_2)_4{-}CH_3$	18:2 or $C_{18}\,\Delta^{9,12}$
Linolenic acid	$HOOC{-}(CH_2)_7{-}CH{=}CH{-}CH_2{-}CH{=}CH{-}CH_2{-}CH{=}CH{-}CH_2{-}CH_3$	18:3 or $C_{18}\,\Delta^{9,12,15}$
Arachidonic acid	$HOOC{-}(CH_2)_3{-}CH{=}CH{-}CH_2{-}CH{=}CH{-}CH_2{-}CH{=}CH{-}CH_2{-}CH{=}CH{-}(CH_2)_4{-}CH_3$	20:4 or $C_{20}\,\Delta^{5,8,11,14}$

*Systematic names are simply derived. For examples, lauric acid is *n*-dodecanoic acid and linolenic acid is *n*-hexadecadienoic acid.

†Both carbons of a double bond may be designated, as in $C_{16}\,\Delta^{9:10}$.

The shape of these phosphoglycerides is shown in Figure 2-1, which is a picture of a CPK model[2] of dipalmitoylphosphatidylcholine. It is a very asymmetric molecule with an elongated form. There is a polar group at one end and a relatively long hydrocarbon portion at the other. The polar group is stable when surrounded by water molecules (also polar). Such groups are said to be *hydrophilic*. The hydrocarbon chains are unstable in association with water and are said to be *hydrophobic*. A molecule with such differing parts is called *amphiphilic*. The consequences of this situation are more easily understood if the free energy states are considered.

Free energy[3] is a thermodynamic state variable that may be thought of as the exchangeable energy of the system under the particular conditions of the observation. Rather than deal with the complicated system consisting of molecules in a medium of other molecules, first consider the conceptually simple example of a system that consists of a surface supporting a Ping-Pong ball of mass m that at a given instant occupies any one of three closely spaced shallow depressions 1, 2, 3 at heights $h_1 < h_2 < h_3$. Assume that the free energy of the ball is represented entirely by kinetic energy and gravitational potential energy mgh, where g is the gravitational constant. Let the surface be vibrating so as to impart a time-average rotational and translational kinetic energy K to the ball, thereby introducing the tendency to change position (energy) from one depression to another. The probability P_i of seeing the ball in the ith depression on a random instantaneous look is

$$P_i = \frac{\exp(-mgh_i/K)}{\exp(-mgh_1/K) + \exp(-mgh_2/K) + \exp(-mgh_3/K)}$$

The probability P_i may be interpreted alternatively as the fraction of time (any sufficiently long time; infinity to be precise) the ball spends in the ith depression. If K is very much greater than mgh_i for all i, then the ball is equally likely to be in any depression. For smaller K and $h_3 \gg h_2 \cong h_1$, the ball almost never occupies depression 3. If $h_1 \ll h_2 < h_3$ and K is not too great, then the system is said to be *stable* in depression 1, and the free

[2] CPK (Corey-Pauling-Koltun) models are scaled representations of bond angles and van der Waals radii of atoms associated by covalent bonds. They are useful representations of the space-occupying features of molecules.

[3] There are various forms of free energy, depending on the modes of energy content and exchange and the descriptive variables of a particular process. In all cases the free energy represents energy recoverable as work. See Annotated Bibliography, Thermodynamics, for references.

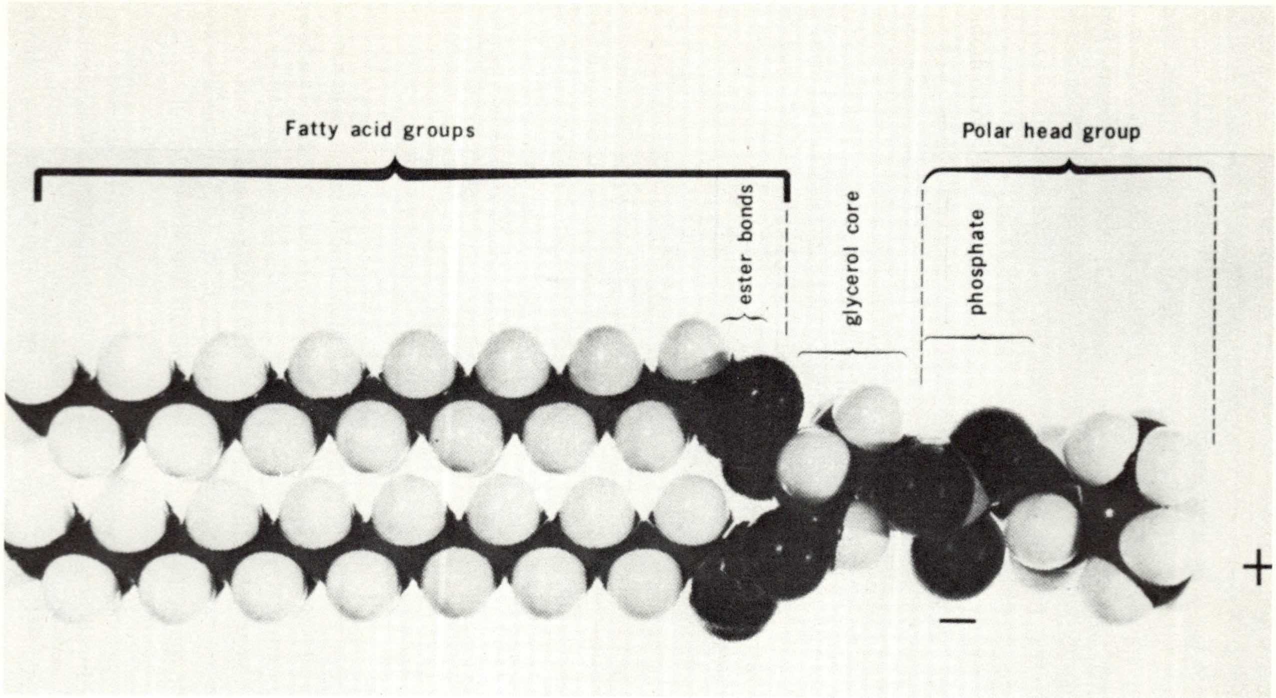

Figure 2-1 Spatial relations of parts of the choline phosphoglyceride molecule. The three-carbon glycerol core links a hydrophilic phosphate with two hydrophobic fatty acid groups. This molecule is dipalmitoylcholine phosphoglyceride. The fatty acid chains are saturated and in the all-trans state (see p. 23). Photograph of a CPK model (see footnote 2, p. 13).

energy change required to escape depression 1 is large relative to K, the available randomly exchangeable energy.

In the case of molecules of phosphoglycerides in water the consequences of their amphiphilic nature are easily understood by using this sort of analysis. The free energy of a system of phosphoglyceride molecules and water molecules is determined by polar and hydrophobic interactions (corresponding to gravitational energy mgh in the above example). The water molecules and the polar head groups interact through electrostatic and electrodynamic forces so that they are more stable in association with one another than they are in association with the nonpolar hydrophobic tails of the phosphoglycerides. Thus, the free energy is reduced by excluding water from between hydrophobic chains and by having polar heads associated with one another and with water.[4] This lower energy state is the stable state for such a system. The randomly exchangeable energy that makes shifts from state to state possible (corresponding to K in the example above) is the thermal energy kT, which is proportional to the temperature (absolute) of the system (k is the Boltzmann constant). As a result, phosphoglycerides–as well as similar asymmetric, amphiphilic molecules–form *micelles* of various forms in water. Some of these are illustrated in Figure 2-2.

A physically similar molecule, sphingomyelin, is also a major component of human erythrocyte membrane lipid. It is about as abundant as choline phosphoglyceride and ethanolamine phosphoglyceride; the three together account for 83 mol% of the phospholipids. The chemical structure is shown in Figure 2-3. Sphingomyelin is a sphingolipid, a class of compounds with similar structure but varying in chain length of the hydrophobic carbohydrate portion and in the monophosphate ester polar group. (Branched forms occur among bacteria; the unbranched forms occur in animals.) The differences from the phosphoglycerides are seen in the structures (Figures 2-1 and 2-3). For the purpose of understanding biomembranes, however, the similarities are significant. Sphingomyelin, like a phosphoglyceride, is an elongated molecule with a polar, hydrophilic head and a long hydrophobic tail. It is amphiphilic and forms micelles in aqueous media. Like choline phosphoglyceride, it is a net neutral zwitterion.

Cholesterol is the major lipid component of the red cell membrane

[4] Entropy makes a contribution to free energy. It is not dealt with explicitly here. A complete treatment of this problem for phosphoglycerides is given in C. Tanford, "The Hydrophobic Effect: Formation of Micelles and Biological Membranes," Wiley, Toronto, 1973. One may think of the increase of order (decreasing entropy), as in the lining up of the phosphoglycerides, as requiring energy. The stable state described here loses more energy (enthalpy) in the association than the energy gain corresponding to the decrease in entropy.

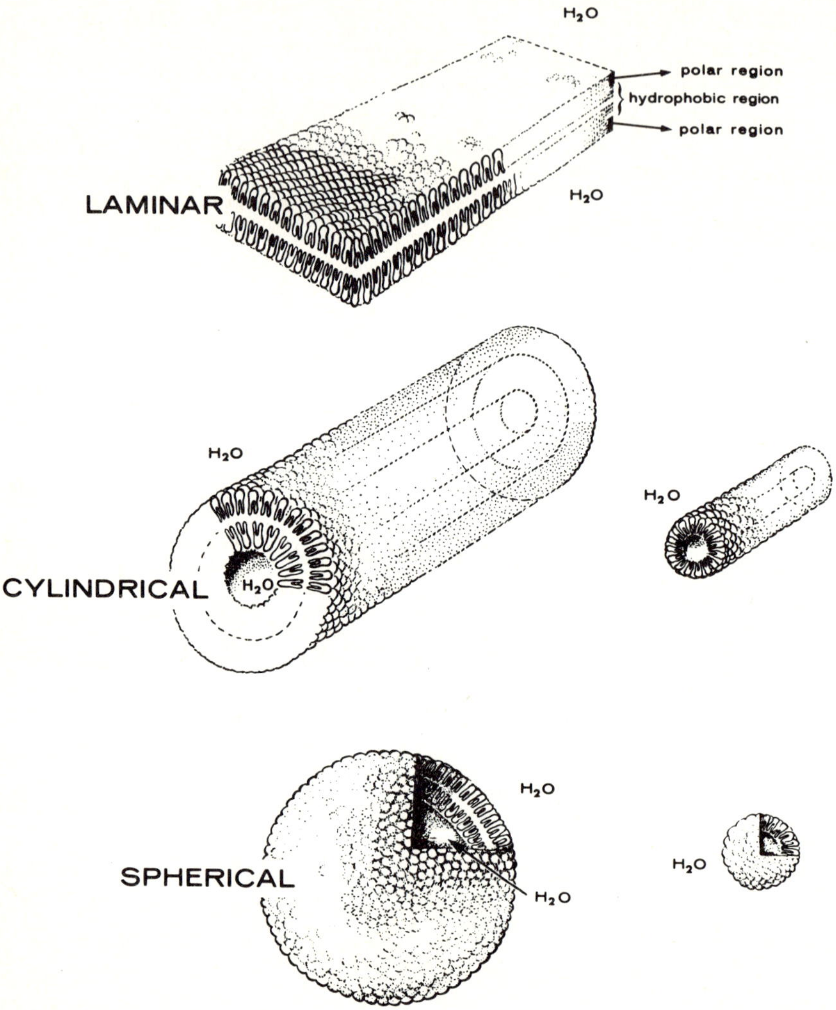

Figure 2-2 Forms of micelles of phospholipids in water. The configurations all decrease contact between hydrophobic regions of the molecules on the one hand and water or polar portions on the other hand. Note that the spherical and cylindrical forms may be single-layered (not containing water) or double-layered (containing water).

remaining to be discussed. It is 25% by mass of the total lipids of the human erythrocyte membrane; usually it accounts for 40-50 mol% of the membrane lipid. Cholesterol is generally a major lipid component of plasma membranes and is found in other cell membranes as well, although it is a minor component of, or missing from, some "deeper" membranes such

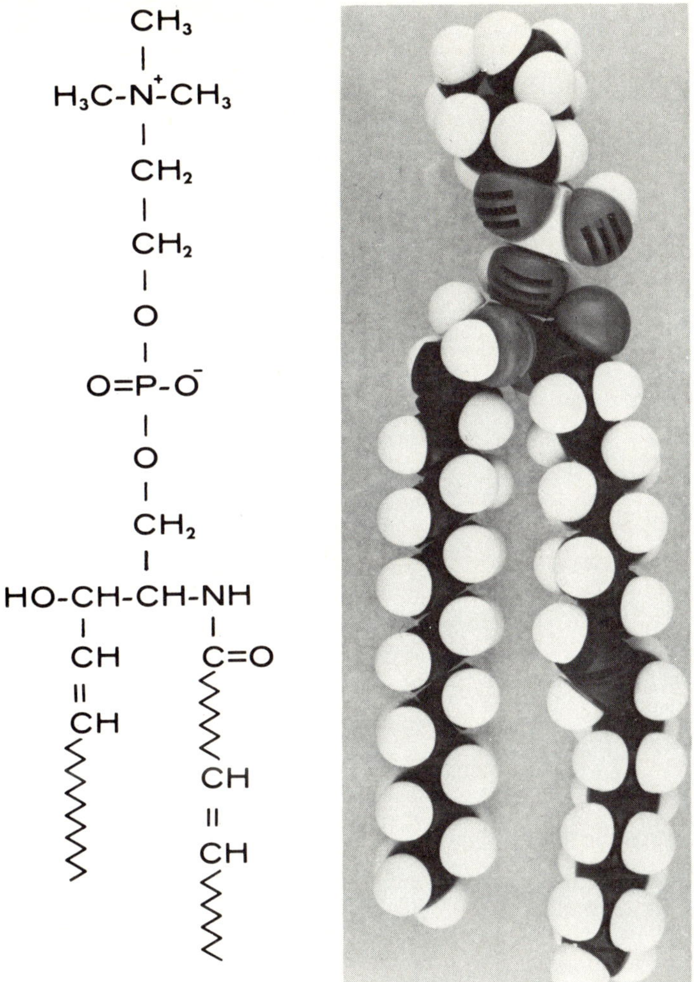

Figure 2-3 Chemical structure and CPK model of the sphingomyelin molecule. The angled lines represent hydrocarbon chains; each line represents a bond, each vertex a carbon atom (hydrogen atoms not shown). Although the linkage between the polar choline phosphate group and the hydrophobic hydrocarbon chains is different, the overall configuration and dimensions of the molecule are similar to those of the phosphoglycerides. Compare with Figure 2-1.

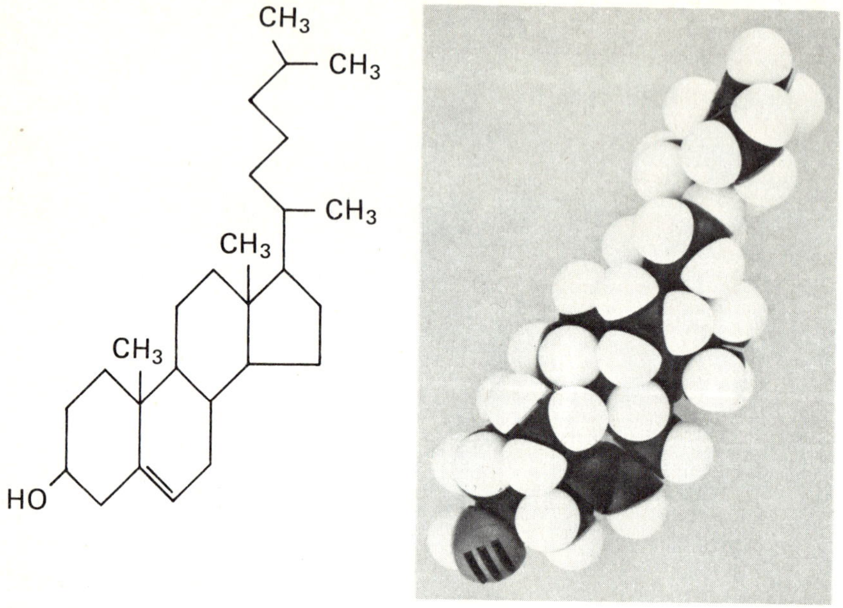

Figure 2-4 Chemical structure and CPK model of cholesterol. The hydroxyl group is polar and, with the hydrophobic side chain of the five-carbon ring, forms an elongated amphiphilic molecule.

as certain nuclear membranes. Cholesterol is a sterol; its structure and CPK representation are given in Figure 2-4. It is a shorter molecule than the phosphoglycerides and sphingolipids with 16- or 18-carbon hydrocarbon chains. However, like these other lipids, it is amphiphilic, most of its surface being hydrophobic. The hydroxyl group on carbon 3 of ring A makes the molecule polar at one end of its somewhat elongated shape. This hydroxyl is capable of hydrogen bonding.

Cholesterol and phospholipids are virtually insoluble in water. Phospholipids are almost exclusively associated with membranes, but cholesterol is also found stored in animals as cholesterol and cholesterol esters. It is a major component of lipoproteins in transit from sites of absorption and synthesis to sites of incorporation into membranes and fat stores.

The major lipids of the human red cell membrane, in summary, have an approximate relative abundance given by the following values in mol%: cholesterol, 40; phosphatidylcholine, 17; sphingomyelin, 17; phosphatidylethanolamine, 16; phosphatidylserine, 8; phosphatidic acid, 2; and phosphatidylinositol, 1.

There are other lipid components of biomembranes that are usually minor components. These include plasmalogens, which have an unsaturated ether linkage between glycerol carbon 3 and a long-chain hydrocarbon rather than the more common ester linkage, and other variations of the phosphoglyceride form. There are also variations on the sphingolipid theme. A class called glycolipids also appears in association with membranes; these compounds have in common a hydrophilic carbohydrate head group (often galactose or glucose) and hydrophobic tails of either the sphingosine or diacylglycerol type. Again the essential point is that these are all elongated molecules with long hydrophobic tails and polar heads.

A particularly interesting variant of phosphoglycerides occurs when the ester linkage of glycerol carbon 2 is hydrolyzed and the 1-monoacyl derivative is formed. Such a compound formed from choline phosphoglyceride (lecithin) is called lysophosphatidylcholine (lysolecithin). There are enzymes that catalyze the formation of lysolecithin from lecithin. Lysolecithin seems to destabilize membranes and can in fact cause lysis of cells, presumably when sufficient amounts are either incorporated into the plasma membrane or replace lecithin. Phospholipase A enzymes, which catalyze hydrolysis at carbon 2 of the glycerol moiety, are components of some hemolytic venoms.

2-2 AMPHIPHILIC MONOLAYERS

The physical characteristics of membrane lipid molecules produce effects that are central factors in membrane structure. These effects are largely revealed in the behavior of monolayers of the molecules. A monolayer is a three-dimensional array one molecule thick. Monolayers may be formed on the surface of a solid in a gas, liquid, or vacuum; between immiscible liquids; or between a liquid and a gas. Being amphiphilic, phospholipids form monolayers at an air-water interface. The polar heads associate with water, and the hydrophobic tails rise side by side toward the air. Cholesterol also forms a monolayer on water. Figure 2-5B shows the spatial relations of a phospholipid monolayer.

The apparatus represented in Figure 2-5A can be used to study various properties of a monolayer at an air-water interface. It is possible to measure the difference in surface tension (force per unit length) at the air-water interface with and without the monolayer; this difference is called the surface pressure π. The bar across the trough at the interface can compress the monolayer. The measured force on the bar divided by the length across the trough gives the surface pressure. Knowing the amount of phospholipid

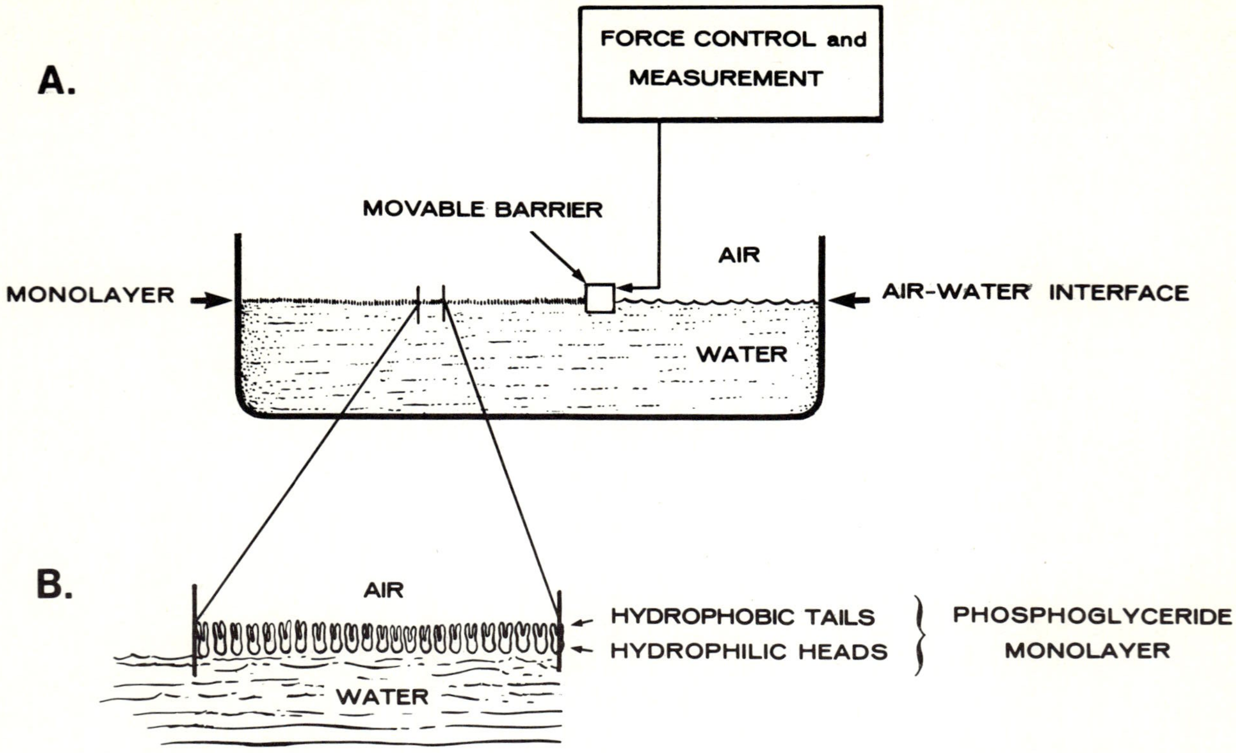

Figure 2-5 Schematic representation of an apparatus for monolayer experiments. (A) Measurement of relation between surface pressure and area of monolayer. (B) Detail indicating orientation of molecules in a phosphoglyceride monolayer.

placed on the surface of the water and the area available to it, one can calculate the area per molecule A_m in the monolayer. Figure 2-6 gives the relation between these variables for a monolayer of pure choline phosphoglyceride. The area per molecule decreases as the surface pressure increases. At the highest and lowest temperatures shown, the choline phosphoglyceride monolayer behaves as a two-dimensional gas and a compressible liquid, respectively. Such an analogy suggests that the πA_m relation would vary with temperature. And it does, as shown. At two of the temperatures represented in Figure 2-6, the πA_m relation shows a cusp in an otherwise smooth curve. This discontinuity in behavior occurs at a point (pair of values of π and A_m) characteristic of a given phospholipid and temperature. To the right of the cusp, at high A_m and low π, the monolayer is an expanded liquid film. The angle of the long axis of a particular molecule to the plane of the monolayer varies with time, and the energy of interaction between molecules fluctuates widely. As π is increased from a vanishingly small

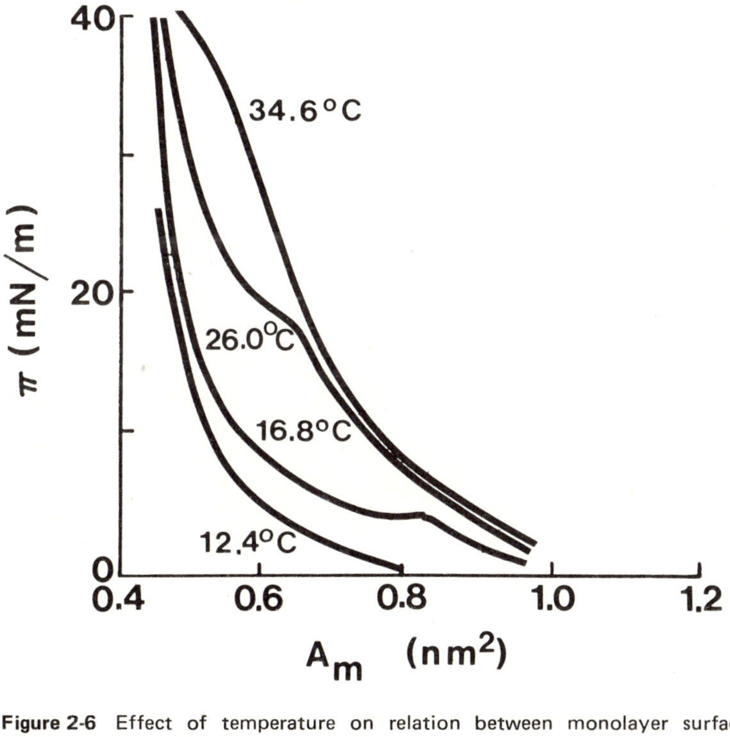

Figure 2-6 Effect of temperature on relation between monolayer surface pressure and area per molecule. These curves represent the behavior of 1,2-dipalmitoylcholine phosphoglyceride at the interface between air and 0.1 *M* NaCl in H_2O. (Adapted from M. C. Phillips and D. Chapman, *Biochim. Biophys. Acta* 163: 301, 1968.)

value, A_m decreases, more for a given change in π at higher A_m. The increasing surface pressure restricts the freedom of motion of a molecule. To the left of the cusp in the curve, the monolayer is in a condensed state. Where the curve is very steep, the molecules are aligned and closely packed and can yield only slightly, if at all, to increasing π. The transition between these portions of the curve—at the cusp—is where the monolayer is in transition between liquid and crystalline forms. That the location of this characteristic point is sensitive to temperature is to be expected. The temperature of the monolayer determines the translational and rotational motions of the molecules and the groups within the molecules. These motions, increasing with increasing temperature, determine the time-average size of the molecule so that the greater the temperature the more area of the monolayer is occupied by a single molecule. We may think of the temperature of this liquid-to-crystal transformation for a given phospholipid as characteristic of a given value of π or, conversely, we may think of the value of π at a cusp on the πA_m curve as characteristic of given temperature for that phospholipid.[5]

However we look at it, the behavior of a phosphoglyceride in a monolayer with respect to the πA_m relation can be separately related to the characteristics of the polar head group and the fatty acid residues. Figure 2-7 shows such differences. For a given fatty acid composition, choline phosphoglyceride occupies a smaller area per molecule at a given surface pressure than does ethanolamine phosphoglyceride. The monolayers from which the data of Figure 2-7 were obtained were formed on sodium chloride (NaCl) solutions, so the ethanolamine phosphate groups are largely negative and the choline phosphate groups are largely neutral zwitterions. One expects an electrostatic expansion for ethanolamine phosphoglyceride relative to choline phosphoglyceride. The curves for the dimyristoyl (C_{14}) phosphoglyceride show a difference in form as well as a simple displacement of the πA_m curves. The curve for the choline form shows a cusp related to the liquid-crystal transformation, while that for the ethanolamine form is smooth throughout. Possibly the electrostatic forces permit crystallization of a lecithin monolayer at 22°C but prevent such crystallization for ethanolamine phosphoglyceride.

The effects of chain length are also exhibited in Figure 2-7. For these saturated forms, increasing chain length displaces the curves to the left—that is, for a given π, A_m decreases as chain length increases. This reflects the hydrophobic interaction that increases the intermolecular attraction

[5] A quantitative analysis of this system follows the basic notion of the gravitational example on p. 13. Here thermal energy kT corresponds to K of that example, and the energy of electrostatic, electrodynamic, and entropic interactions corresponds to mgh.

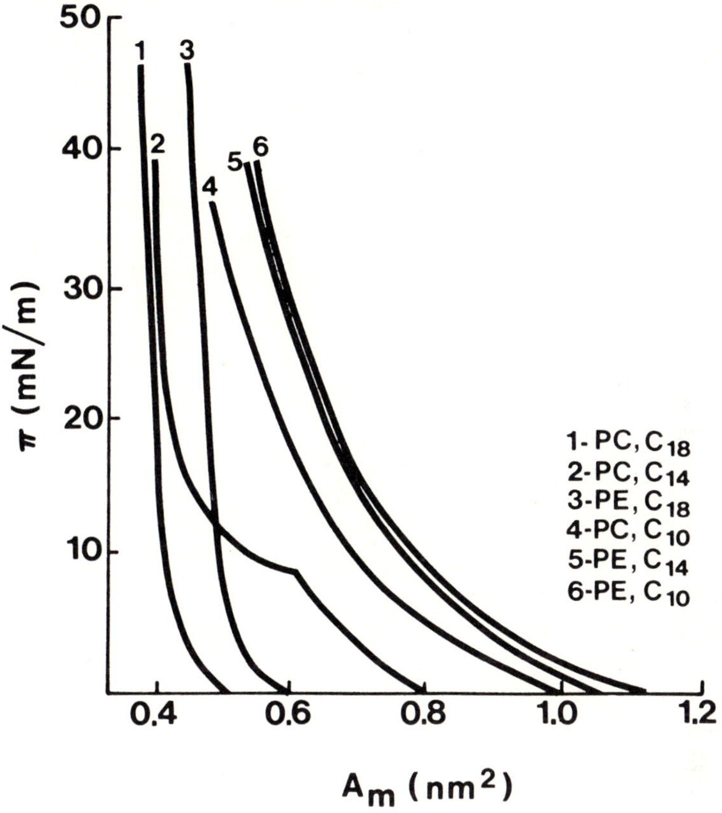

Figure 2-7 Effects of head groups and fatty acid groups on phosphoglyceride monolayer behavior. PC, choline phosphoglyceride; PE, ethanolamine phosphoglyceride; C_n, diacylated with saturated fatty acid of *n* carbons. (Adapted from M. C. Phillips and D. Chapman, *Biochim. Biophys. Acta* 163: 301, 1968.)

(in the presence of H_2O) as chain length increases. In addition to variation in length, the hydrocarbon chains can vary in degree of saturation. The presence of a single double bond produces the considerable change in monolayer behavior shown in Figure 2-8. Stearoyloleoylcholine phosphoglyceride has one double bond between carbons 9 and 10 of one of the fatty acids. A double bond restricts rotation so that the carbon atoms bonded to the adjacent ethylenic carbons are fixed in either a cis or trans position, as represented in Figure 2-9. In either configuration the double bond puts a permanent kink in the hydrocarbon chain, resulting in a greater A_m at a given π and temperature. The monolayer structure may be said to be looser with the double bond in the hydrophobic chain.

Mixtures of amphiphilic elongated molecules also form monolayers at an air-water interface. As might be expected, the behavior of a mixture can be more complex than that of a single pure compound. The effect of changing the molar ratio of a mixture of cholesterol and choline phosphoglyceride, two prominent membrane lipids, is shown in Figure 2-10, in which the average molecular area A_m is plotted against mole fraction of cholesterol in the monolayer mixture. The dotted straight line gives the expected variation in A_m derived from molecules of both types, if the molecules do not interact or intertwine in any special way. The solid line represents the results of experimentally determining A_m for a varying molar ratio. Such data reveals a *condensing effect* of cholesterol, since the observed data points fall below the straight line. That is, the mixed molecules occupy a smaller average area per molecule than the weighted average for pure monolayers.

Not surprisingly, the degree of condensation by cholesterol varies with the phospholipid species at a constant temperature. For lecithins of the

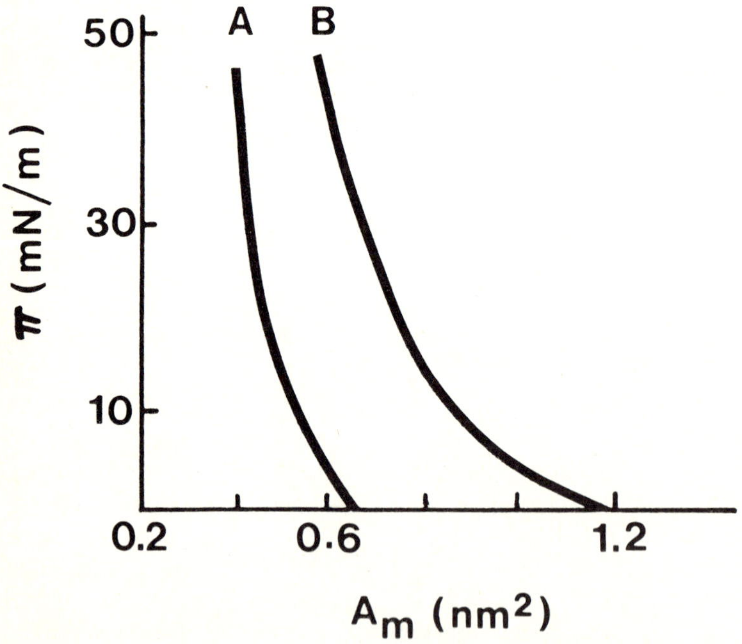

Figure 2-8 Effect of double bond in hydrocarbon chain on monolayer behavior. (A) Curve showing relation to surface pressure to area per molecule for saturated distearoylcholine phosphoglyceride. (B) Curve for unsaturated stearolyoleoylcholine phosphoglyceride. The double bond is associated with "looser" packing.

series, dicaproyl (C_{10}), dilauroyl (C_{12}), dimyristoyl (C_{14}), dipalmitoyl (C_{16}), distearoyl (C_{18}), and dibehenoyl (C_{22}), at 22°C and 5 mN/m surface pressure, the effect is maximum for the dipalmitoyl form. The condensation is greatest for all forms near mole fractions between 0.25 and 0.75. For longer chain lengths the hydrophobic interaction is so great at this temperature that the monolayer is quite compact without cholesterol. For the C_{22} form the data

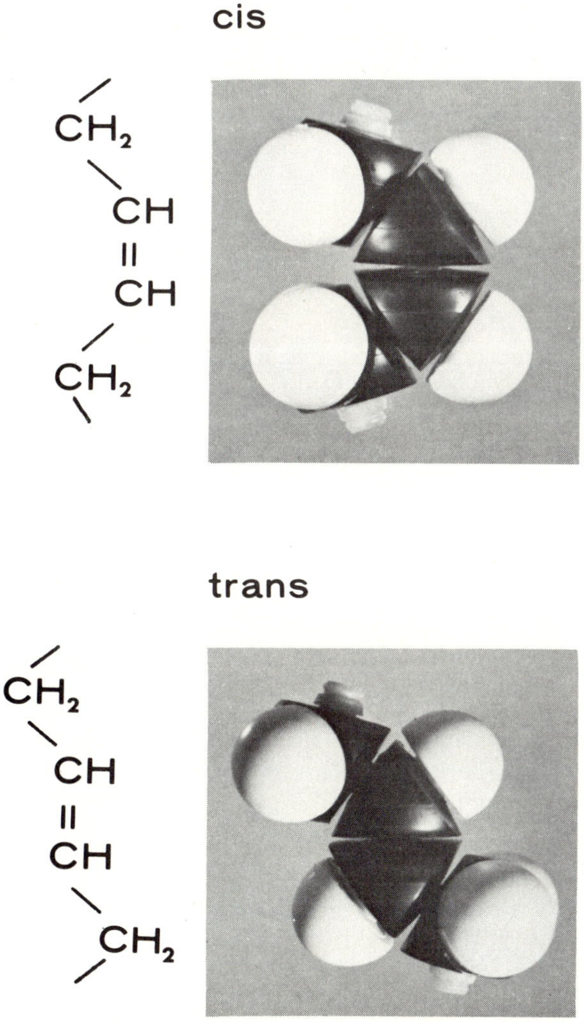

Figure 2-9 Chemical structures and CPK models of the region including a double bond in a hydrocarbon chain.

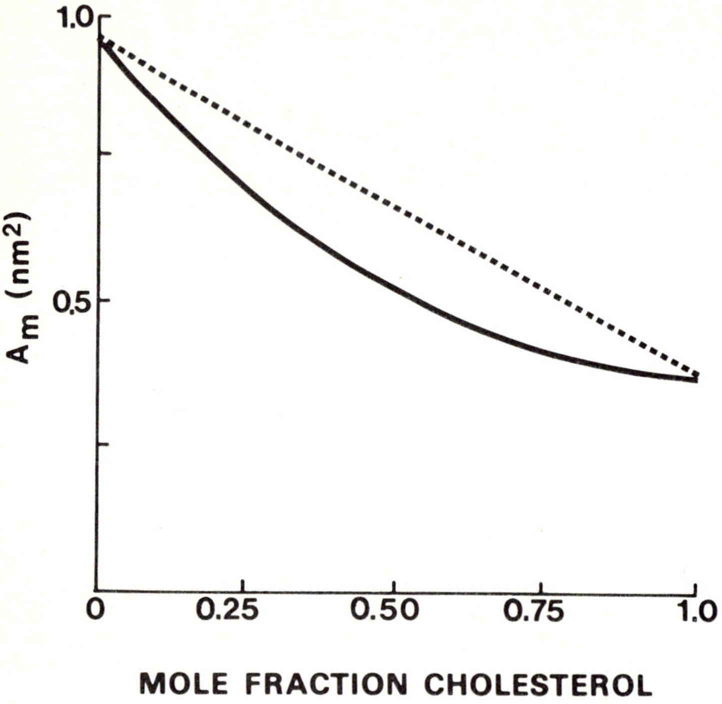

Figure 2-10 The condensation effect in a mixed monolayer of cholesterol and lecithin. The dotted line gives the theoretical area of separate pure phases; the solid line gives the experimental area of the mixture.

points fall on the straight line between pure cholesterol and pure dibehenoyl lecithin; no condensation occurs. The same series at 22°C and 20 mN/m (surface pressure four times greater) shows a maximum condensation for the C_{10}, C_{12}, and C_{14} forms. The exact mechanism of this phenomenon is not known with certainty, but two kinds of effects may be anticipated. On the one hand a simple steric effect might result from partial packing of cholesterol molecules into the interstices among the choline phosphoglyceride molecules. This is consistent with the effects of changing surface pressure and number of double bonds. On the other hand the electrostatic, electrodynamic, and entropic changes produced by the mixing may be responsible for condensation. These effects are not exclusive of one another, so for now we may accept the observations as a result of some combination of mechanisms. The various interactions of molecules in a monolayer provide a rich conceptualization for understanding such arrays.

In addition, the interaction of a monolayer with adjacent bulk phases is also of interest. At equilibrium there is an exchange, between monolayer and bulk phase, of components such as water molecules, ions, and small solutes. The equilibria for such exchangeable components are affected by the surface pressure and temperature of the monolayer as well as by its composition. Clearly the free energies of exchangeable ions and molecules in the monolayer are dependent on surface pressure and intermolecular forces.

2-3 MEMBRANE PROTEINS

As a prelude to the discussion of membrane proteins, clarification of terminology is in order. The fundamental defining characteristic of a protein is the long peptide chain. By "long" is meant greater than, say, 100 amino acids. The sequence of polymerized amino acids determines the *primary structure* of the protein. The polypeptide undergoes rotations of peptide bonds and interactions between neighboring amino acid units, thus determining the *secondary structure.* The resulting sequence of helical or sheetlike portions of the polymer in turn determines the folding and disulfide bond formation (between cysteine components) that produce the *tertiary structure.* Such proteins sometimes interact between or among themselves or with other protein "monomers" to form *oligomeric proteins* with a *quaternary structure.* It has become common usage to refer to such complexes as dimers, tetramers, and so on, although the terms refer to mostly noncovalent bonding between polypeptides that are themselves polymers of amino acids. Such complexes of polypeptides are formed by disulfide bonds, hydrogen bonds, ionic bonding, and more nonspecific interactions such as the van der Waals force and the hydrophobic effect described in the discussion of amphiphilic lipids.

The proteins of membranes may be classified in various ways. From the viewpoint of the experimenter, one obvious way is to classify a protein according to what one must do to isolate it from the rest of the membrane. This approach has given us the two classes: *intrinsic,* or *integral, protein* and *extrinsic,* or *peripheral, protein.* One should bear in mind here the matters discussed in Section 1-3—the problem of defining the biomembrane. The extrinsic membrane proteins are separable by chemically less disruptive measures; reducing the ionic strength or altering the H^+ concentration of the aqueous suspending medium will do. The intrinsic proteins remain associated with the lipid and with the morphological entity under such mild treatment; they are separated only by the application of agents, such as detergents, that interfere intimately with intermolecular interactions.

Other ways of classifying the proteins of biomembranes are less directly related to the proteins' being parts of the membrane. There are the two classes defined by tertiary structure: globular and filamentous. These are self-explanatory geometric terms. There are the classes defined by chemical content: pure peptides, glycoproteins (containing carbohydrate groups), and lipoproteins (containing lipid). In addition there are the functional classes: enzymes and structural proteins. This last classification is to be discouraged, as there are proteins that have been considered structural and seem to determine morphological characteristics, yet that catalyze certain hydrolytic reactions when isolated. There are also structural features of membranes that can be attributed to proteins with specific catalytic activity or other clearly functional properties. The origins, advantages, and disadvantages of these various systems of classification become clear as one looks at biomembranes in detail. The terms are introduced here since they are so often seen in journals and books.

Membrane proteins, like membrane lipids, vary with the type of membrane, type of cell, and species. One may expect greater variation in proteins since the complex structure of proteins is a major mode for expressing specificity in biologic systems. In particular it is the details of protein structure that determine the many examples of catalysis of different chemical reactions occurring in different tissues, cells, and species. For the purpose of presenting an example in some detail, the human red cell membrane is again useful. The proteins of this membrane have been studied extensively and very successfully. Furthermore, having described the lipids and proteins of the same membrane, one is able to get a detailed picture of the molecular organization of a particular biomembrane.

The cytoplasm of the erythrocyte is an apparently homogeneous aqueous solution of hemoglobin, the usual intracellular ions, metabolic substrates, and enzymes. The plasma membrane can be made permeable to all these contents (a process called hemolysis) in a controlled manner to produce red cell *ghosts.* The ghosts can be separated by centrifugation and washed under conditions that retain all membrane components. Ghosts are widely considered to be pure plasma membrane. This procedure is not necessary for separation of lipids because extraction with a lipid solvent immiscible with water leaves the proteins of the membrane in the aqueous phase or precipitated, along with cytoplasmic cell components. The separation of membrane proteins, however, must be preceded by isolating the membrane from other cell components, and the choice of procedure and interpretation of results must be made with the possibility of the binding of cytoplasmic components in mind. In the case of extrinsic proteins there are morphological correlations (presented in Section 2-4) that support the conventional view that these proteins are genuinely a part of the red cell membrane. Peripheral and integral proteins together account for about half the mass of the human red cell membrane; lipids account for the other half.

The extrinsic proteins are removed from the membrane by a change of ionic strength or change of H^+ concentration, which leaves the intrinsic proteins associated with the lipids in a morphologically characteristic structure that represents the greater part of the membrane. In the case of the erythrocyte membrane the extrinsic proteins, so defined, constitute some 25-40% (by mass) of the protein. This is atypically high as membranes go. The intrinsic proteins can be removed from the morphological structure—and separated from the lipids—with variable efficacy by reagents that interfere with hydrophobic interactions, such as detergents, and by reagents that disrupt hydrogen bonds. These reagents often denature proteins, adding to the difficulty of characterizing them once they are isolated.

The method most often used to separate and characterize proteins of the erythrocyte is sodium dodecyl sulfate (SDS)-polyacrylamide gel electrophoresis. A brief discussion of this method and what one finds with it will serve to illustrate the important properties of the various proteins of the red cell. Electrophoresis is the migration of a material through an electric potential gradient. The driving force for such a motion is the interaction between the electric potential and the charge on the moving body. This force is proportional to the net charge on the protein in this case. The supporting medium for the molecules is an aqueous phase in the gel formed by the wet polymer, polyacrylamide. This medium offers resistance to the moving molecule or aggregate of molecules such that the larger the molecule or aggregate the greater the resistance. It is possible on the basis of some knowledge of the material being examined and by using well-understood marker molecules to estimate molecular masses or masses of molecular complexes by gel electrophoresis.

SDS is an anionic detergent (Figure 2-11), which is effective in freeing intrinsic proteins and associates with proteins in a rather consistent way. It disrupts tertiary structure so as to make peptide chains more or less linear, and it binds to the unfolded protein in a constant mass ratio (mass SDS/mass protein). This binding masks the electric charges on the proteins and imbues the complex of protein and SDS with a charge proportional to the amount of bound SDS and therefore proportional to the molecular mass of the protein. The mobility (final velocity per unit electric potential gradient) of this complex of SDS and peptide is approximately inversely proportional to the logarithm of the molecular mass. There is much to be cautious about in the interpretation of the results of SDS-polyacrylamide gel electrophoresis applied to membrane proteins, but it has been very effective in a variety of ways. When the gel is stained to reveal

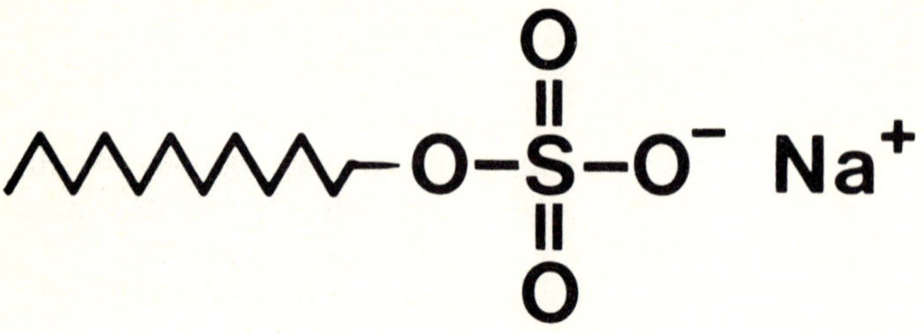

Figure 2-11 Chemical structure of SDS.

the protein after migration, the optical density of the stained gel can be determined and plotted against distance along the gel. Such a representation is shown in Figure 2-12, where the estimated molecular masses are associated with the major peaks. We will see that these peaks and masses are not all to be taken as representative of single species of monomeric proteins.

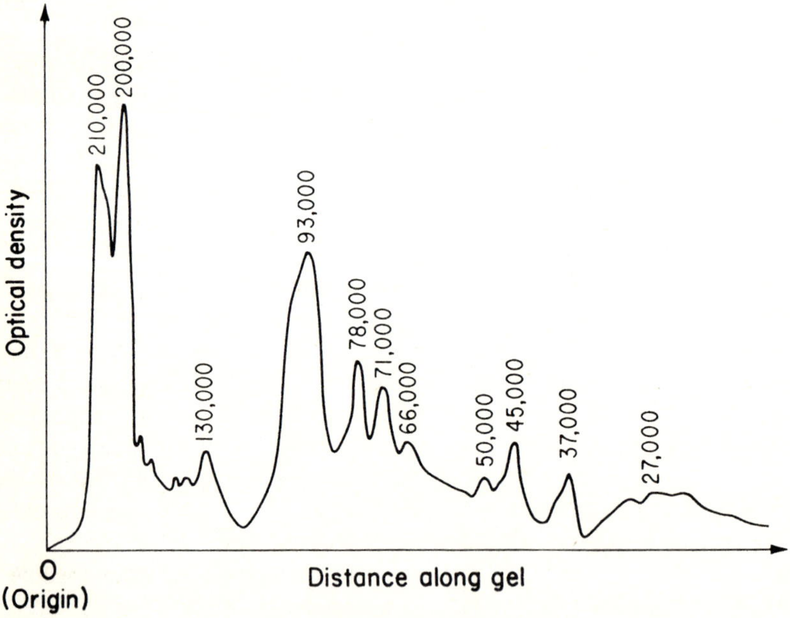

Figure 2-12 Electrophoretic pattern of proteins extracted from human erythrocyte membrane. The curve represents the optical density along a column of polyacrylamide gel stained for protein after electrophoresis. The protein was extracted with SDS. Numbers at the major peaks are estimates of molecular weights in daltons. (From H. R. Trayer, Y. Nozaki, J. A. Reynolds, and C. Tanford, *J. Biol. Chem.* 246: 4485, 1971.)

The above description of the SDS-gel electrophoresis method holds little promise for characterizing proteins that include covalently bound groups other than amino acids in a peptide chain. Such nonpeptide groups will not bind SDS in the same way as do peptides. One of the bands revealed on the gel by peptide stain does contain carbohydrates: band 3[6] (or component 3) is the major intrinsic protein, constituting some 30% of this type of protein. It contains between 5 and 10% carbohydrate–mostly mannose, galactose, and *N*-acetylglucosamine, probably glycosidically bonded to asparagine residues. (Some authors do not class band 3 as a glycoprotein, reserving the term for proteins with much higher proportions of carbohydrate.) The small portion of carbohydrate does not prevent staining and permits an estimate of molecular mass; band 3 migrates as if it were a protein of about 90,000 daltons. For other glycoproteins the situation is different.

If, after electrophoresis of human erythrocyte membrane proteins, one stains a polyacrylamide gel with periodic acid-Schiff (PAS) reagent (which oxidizes the hydroxyl groups of carbohydrate and binds a stain to the resulting aldehydes), four or more bands that were not revealed by the protein stain appear. How many unique glycoproteins are represented by these bands is not known with certainty. It is relatively certain, however, that in mammals they are characteristic of plasma membranes and membranes of the Golgi apparatus and are absent, or very nearly so, from other types of biomembrane. The carbohydrates of these glycoproteins account for much of the molecular mass. One of them, *glycophorin,* has been thoroughly studied. It is a single peptide of about 130 amino acids with some 16 oligosaccharide chains–all covalently bonded to serine and threonine residues, except for one chain bonded to asparagine. The carbohydrate portion is 60% of the mass, and all of it is bonded on the first 50 amino acid residues from the amino-terminal end of the peptide chain. The significance of this structure will be pointed out in the following section.

2-4 SPATIAL ORGANIZATION OF MOLECULES IN BIOMEMBRANES

The first observations of cellular structure presented biologists with a problem: How is the cell maintained as a distinct structure? The possible answers seem to be: Cells exist only in holes in a surrounding matrix; the cell is a gel or solid; and the periphery of a cell is a confining barrier–a

[6] This terminology reflects current usage, derived largely from a paper by G. Fairbanks, T. L. Steck, and D. F. H. Wallach (*Biochemistry* 10: 2606, 1971). Techniques with greater resolution show more bands than others, often with intermediate mobilities.

membrane. The first possibility is applicable to certain plants, but even here the cells can be seen to shrink away from their confining walls under certain conditions. That cells are solids seems untenable in the face of observations of rapid shape changes and swelling and shrinking in response to changes in surrounding media. This sensitivity to concentrations of solutes in surrounding aqueous media (discussed in Chapter 3) can be explained by either a gel model or a membrane model of the cell, but comparison of responses to different solutes was taken by most biologists (at a time when more refined morphological and preparative techniques were unavailable) to favor the last hypothesis. Certainly the early micropuncture experiments and observations of electric properties supported the membrane hypothesis. The results were readily interpreted as showing two volume conductors—extracellular and intracellular—separated by a thin region of high electric resistance.

The existence of the biomembrane was a hypothesis based on the interpretation of such indirect observations until the electron microscope brought morphological methods to a level where the biomembrane could be detected. With this instrument the membrane could be assigned a feasible structure in terms of dimension and continuity. An inhomogeneity within what might reasonably represent the membrane could then be visualized (Figure 2-13). Furthermore, many of the optical discontinuities of cells were found to be bodies enclosed by structures apparently identical to that enclosing the whole cell, the plasma membrane. Electron microscopy suffers the same basic limitations as optical microscopy, or any observation, for that matter. In both electron and light microscopy the image is produced by the influence of interactions between specimen and particles (electrons and photons) on the subsequent interactions between particles and detector. Often such effects must be enhanced or otherwise modified by altering the specimen (with a stain, for example). The electron microscope, by virtue of its greater resolving power, is even further removed from everyday experience where bodies are continuous and behave according to familiar laws of mechanics. One must therefore resist the temptation to carry over the artificial imagery too literally into the conceptualization. Nevertheless, electron microscopy supports two propositions: (1) there are many biomembranes in many cells; and (2) there is a distinct morphological character common to bounding structures within and around cells, and this common character includes inhomogeneity on a molecular scale in the direction perpendicular to the boundary.

The first hint as to the biomembrane's appearance under the electron microscope predates the electron microscope itself. In 1925 Gorter and Grendel reported an observation that is widely assumed to have initiated the modern science of biomembranology. They extracted the lipid of erythrocytes

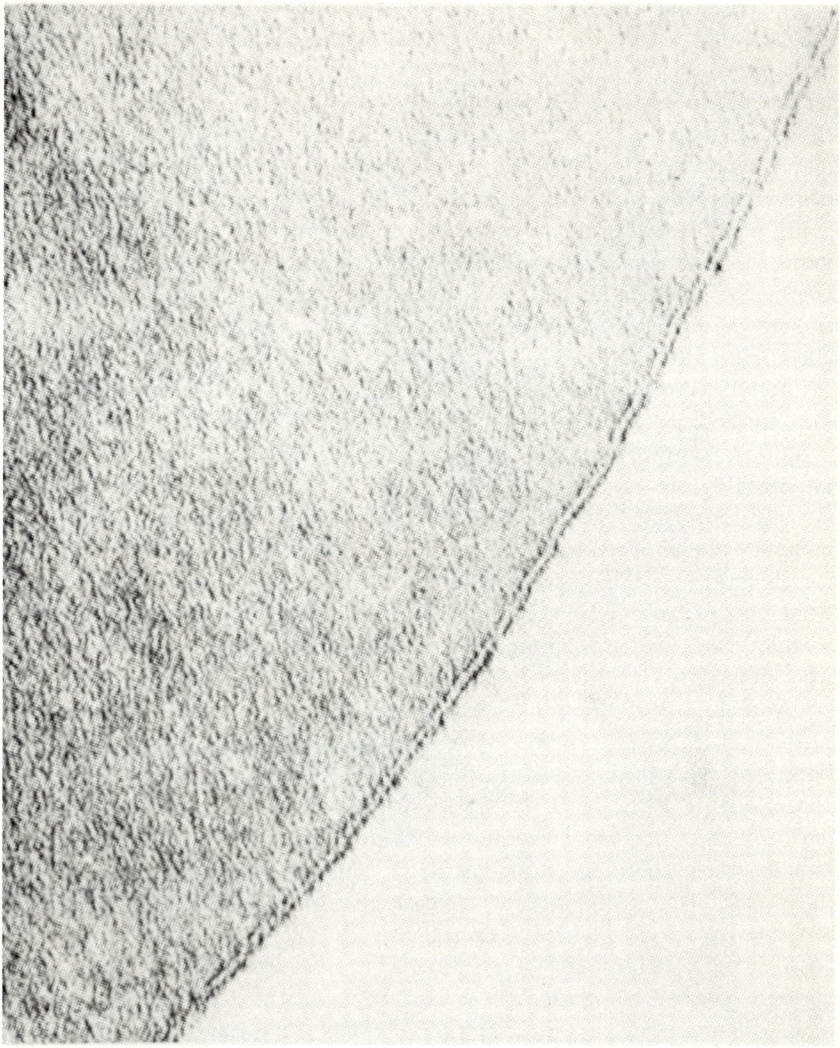

Figure 2-13 Transmission electron microscopic image of a human erythrocyte membrane in cross section. The cell was fixed with glutaraldehyde and osmium tetroxide (OsO_4). The parallel dense lines separated by a relatively unstained line are typical of biomembranes. (From J. D. Robertson, *Arch. Intern. Med.* 129: 202, 1972, Copyright 1972, American Medical Association.)

with acetone and formed a monolayer (as discussed in Section 2-2) that occupied, at a reasonable surface pressure, an area twice the estimated area of the surfaces (membranes) of the cells from which the lipid was extracted. It was concluded that the lipid of the red cell was arranged in a bilayer at the cell surface. In 1935 Davson and Danielli suggested that the lipid bilayer sandwiched between layers of protein is a general form for plasma membranes. This idea was formalized and generalized further by Robertson in the next decade and was called the unit membrane. Much of the investigative work—certainly the chemical and morphological work if not the phenomenological transport and materials science approaches—leading to our present conception of biomembranes has been based on the lipid bilayer model.

Early electron microscopy was in the transmission mode in which images were formed by intercepting (with a phosphor screen or a photographic emulsion) electrons that had passed through very thin—of the order of 10 nm—sections of cells embedded in plastic. Almost all the images that could be consistently recognized as correlates of what was called a biologic membrane had the appearance of Figure 2-13. These were interpreted as sections perpendicular to the plane of the membrane, and the apparent layers of the biomembrane were assumed to be correlates of the two oppositely oriented monolayers of phospholipids and cholesterol or of protein layered over the lipid.

A technique with even greater resolving power was available: X-ray diffraction. X-ray diffraction could be used to measure spacings between regular arrays of atoms with sufficiently different numbers of electrons (the electrons do most of the scattering of X-rays). It was a technique that had been used for many years in physics on relatively convenient materials such as crystals. Biology, too, has its convenient materials, and some were successfully examined by X-ray diffraction. For example, both myelinated nerve and the outer segment of retinal rods have sufficiently orderly and close-packed arrangements of membrane. The results of such examinations can be expressed as relative density of electrons (number per unit volume) in an oriented system as a function of distance through the material in a given direction. Such information is represented graphically in Figure 2-14. It is quite consistent with the morphological inhomogeneity suggested in Figure 2-13. With the chemical data, the Gorter and Grendel results, and the Davson-Danielli hypothesis in mind, one can easily conclude that the biomembrane is at least in part a bilayer of phospholipid with hydrophobic hydrocarbon tails contiguous in the central plane, and with polar groups opposite one another and outside the hydrophobic central region. Such a conceptualization has a satisfying consistency with what is known of micelle formation and monolayer behavior of phospholipids and

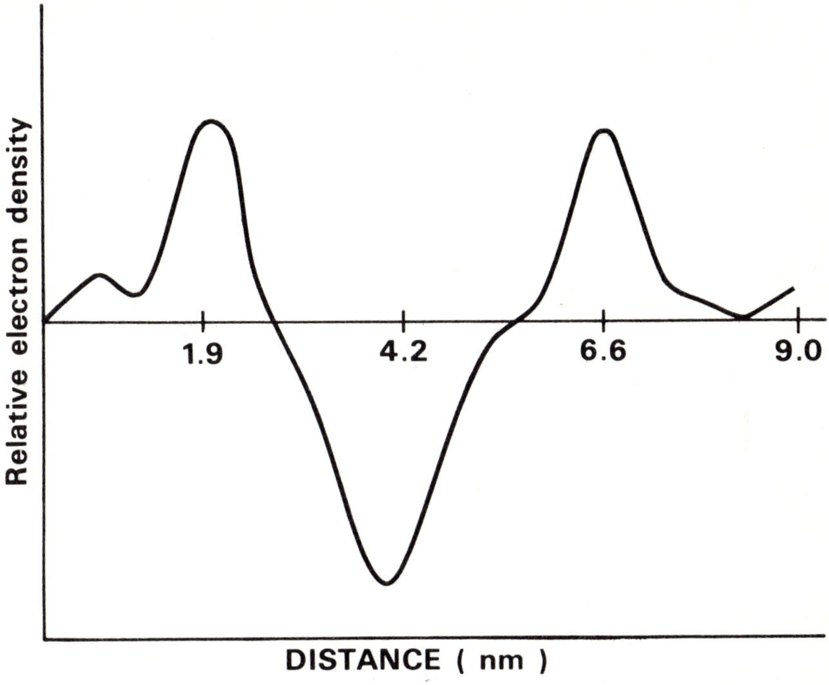

Figure 2-14 Electron volume density across a biomembrane. The curve represents the variation in volume density of electrons relative to water as a function of distance into the membrane (along the normal to the plane of the membrane). This result is for the membrane of axons of the myelinated sciatic nerve of rabbit. (Adapted from D. L. D. Caspar and D. A. Kirshner, *Nature (London) New Biol.* 231: 46, 1971.)

cholesterol. The central organizational feature of a biomembrane, then, is the laminar form of micelle shown in Figure 2-2.

In addition to establishing the general morphological nature of the biomembrane, the electron microscope brought another advantage. It made possible the identification of fractions of cells, separated in various ways, as membranes or containing membranes. In the simplest case, an investigator could hemolyze red cells, centrifuge the preparation, and demonstrate that the sediment was composed entirely of material with the morphological characteristics of a biomembrane, free of hemoglobin and other cell components, which could then be classified as cytoplasmic. In a more complex case, density gradient centrifugation separates a cell into many fractions, several of which can be shown by electron microscopy to be membrane or membrane-bound organelles. In either case, biochemical analysis of the fractions obtained can be correlated with membranes or membrane-bound organelles.

Such separations and morphological correlations give us the chemical content of membranes of various types and show both lipid and protein composition. Lipid extractions, monolayer measurements, and morphological studies support the lipid bilayer hypothesis, but the protein is still to be accounted for. The membrane preparations contain consistent protein fractions but not enough, in most cases, to cover a lipid bilayer on both sides as the early hypothesis suggested. X-ray diffraction and early electron microscopy also did not help place the protein. In one case, however, we can account for protein: there is an apparently loose layer of material, best described as fuzzy, at the cytoplasmic face of the red cell membrane. It is present in both whole cell and ghost preparations where ghosts have been prepared without significant lowering of ionic strength or change in H^+ concentration. However, if ghosts are examined by transmission electron microscopy after procedures that remove what is now called extrinsic membrane protein, this fuzzy layer is not detected. This layer on the cytoplasmic side of the membrane, then, represents most or all of the peripheral protein of the mammalian erythrocyte. Still, there is the problem of how the intrinsic or integral protein is incorporated into the membrane.

There is much evidence that the integral proteins are not simply spread along the polar planes of the biomembrane. The apparent surface tension at a membrane-water interface is not entirely consistent with a protein-water interface; some membranes do not include enough lipid to account for the observed area if the lipid is in a bilayer configuration. Also, the specificities and other peculiarities of transmembrane transport suggest that there is not a simple homogeneous continuous lipid bilayer. These and other factors led a number of investigators to hypothesize that at least some of the membrane proteins interrupt the lipid bilayer and that the membrane is a *mosaic* of lipid and protein in the plane of the membrane.

Morphological confirmation of this hypothesis has been obtained by an extension of electron microscopy that forms an image of a surface formed in the plane of a biomembrane. The method, known as *freeze-fracture* (or freeze-cleaving or freeze-etching), begins with freezing of the specimen. The frozen block is fractured by a sharp edge, thus exposing two surfaces. A heavy metal, usually platinum and/or gold, is applied by evaporation from a low angle to shadow the contours of the fracture surfaces. The shadowed surface is then coated, usually with carbon, and the resulting *replica* is freed of the specimen material. This thin replica can be examined by transmission electron microscopy. The varying deposition of heavy metal produces differences in opacity to electrons that correspond to the contour along the fracture surface through the specimen.

To interpret freeze-fracture electron microscopic images, one must know where the fracture has occurred. The cohesion in the frozen block, of course, varies from point to point on a molecular scale according to variations in the composition of the block. The fracture surface can pass through extracellular medium, across plasma membrane, along plasma membrane, through cytoplasm, and so on. The morphological appearance of membrane regions of such preparations suggests that the fracture surface may follow either interface between membrane and medium or cytoplasm or follow the central surface of the hydrophobic region. Morphological criteria suggest that most of the extensive membrane faces are the result of fracturing between the two laminae of the bilayer (Figure 2-15). Careful analysis of the images and labeling studies confirm the hypothesis that the major fracture surface is that between the phospholipid tails in the hydrophobic region. That relatively simple system, the red blood cell, has been useful in deciding this issue.

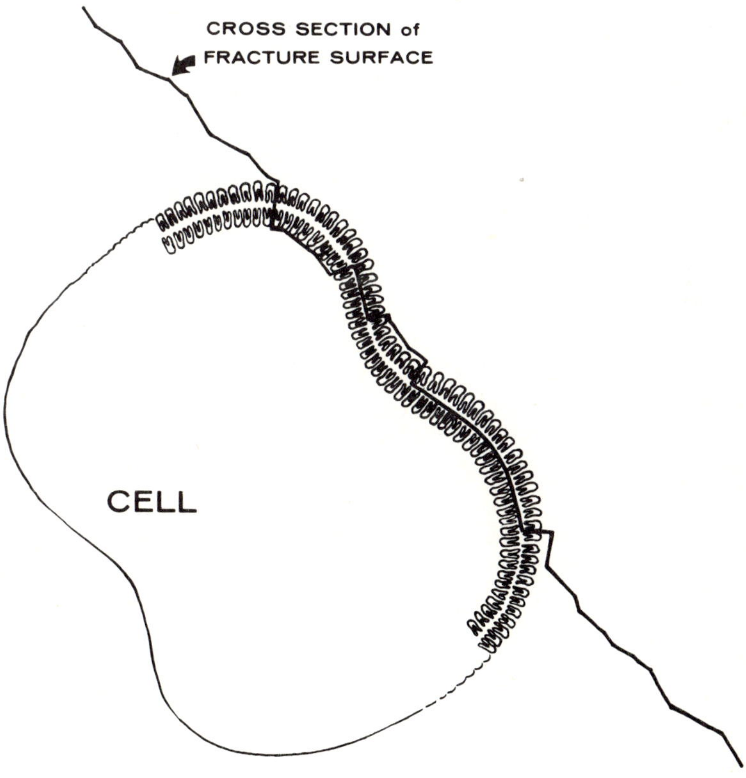

Figure 2-15 Freeze-fracture of a biomembrane: relation of fracture surfaces to parts of the membrane.

However indirect this kind of visualization of biomembranes, it has demonstrated a fundamental characteristic of membrane structure: it is nonhomogeneous in the plane of the membrane. Figure 2-16 illustrates this point. The smooth surfaces of lipid monolayers are interrupted by projections. This feature is characteristic of biomembranes examined by this technique. These particles are often called *membrane-associated particles* (MAP). The number of particles per unit area is different on the two faces: there are fewer on the outer leaflet and more on the leaflet from the cytoplasmic side of the membrane.

Many microscopists first thought of these MAP as being composed of the integral proteins of biomembranes. And indeed there is strong evidence that this is so. The more direct evidence is of two types: ablation and labeling. Treatment of membranes that removes intrinsic protein also reduces the number of particles as visualized by freeze-fracture electron microscopy. Certain proteolytic enzymes alter the number, distribution, and appearance of MAP. Labeling membrane proteins with morphologically identifiable labels (colloidal iron or ferritin-conjugated antibody, for example) has shown a consistent association between aggregation of MAP as seen in freeze-fracture preparations and aggregation of label as seen in thin-sectioned preparations. In addition to this experimental evidence, analysis of the requirements for separation of integral proteins and of the physicochemical nature of the separated proteins suggests that incorporation of protein into the lipid bilayer and extensive hydrophobic interactions of integral proteins are quite likely. It is now widely accepted, on the basis of varied and strong evidence, that the MAP of freeze-fracture preparations are largely, if not entirely, composed of integral membrane proteins.

Much of the evidence that these particles consist of intrinsic proteins was obtained from studies of red blood cells. The morphological evidence suggested that these protein bodies are not equally distributed in the two laminae of the bilayer—or that the interaction with the outer leaflet is more readily disrupted in the freeze-fracture process. In any case this question arises: Do only some or do all the proteins extend through both leaflets of the bilayer? Largely because of the importance of blood transfusion, the human red cell was thoroughly studied with respect to surface antigens long before the problem of location of membrane proteins was recognized. When techniques for separating and identifying membrane constituents were refined, the specific nature of the antigens was a focus of attention. The major blood group antigens were associated with the carbohydrate groups of membrane glycoproteins (glycolipids may be involved in some cases). This constitutes evidence that these groups—at least some of them—are exposed at the outer face of the membrane. Other red cell antigens have been shown to be pure peptides, not glycoproteins, which indicates exposure of peptide chains at the outer surface.

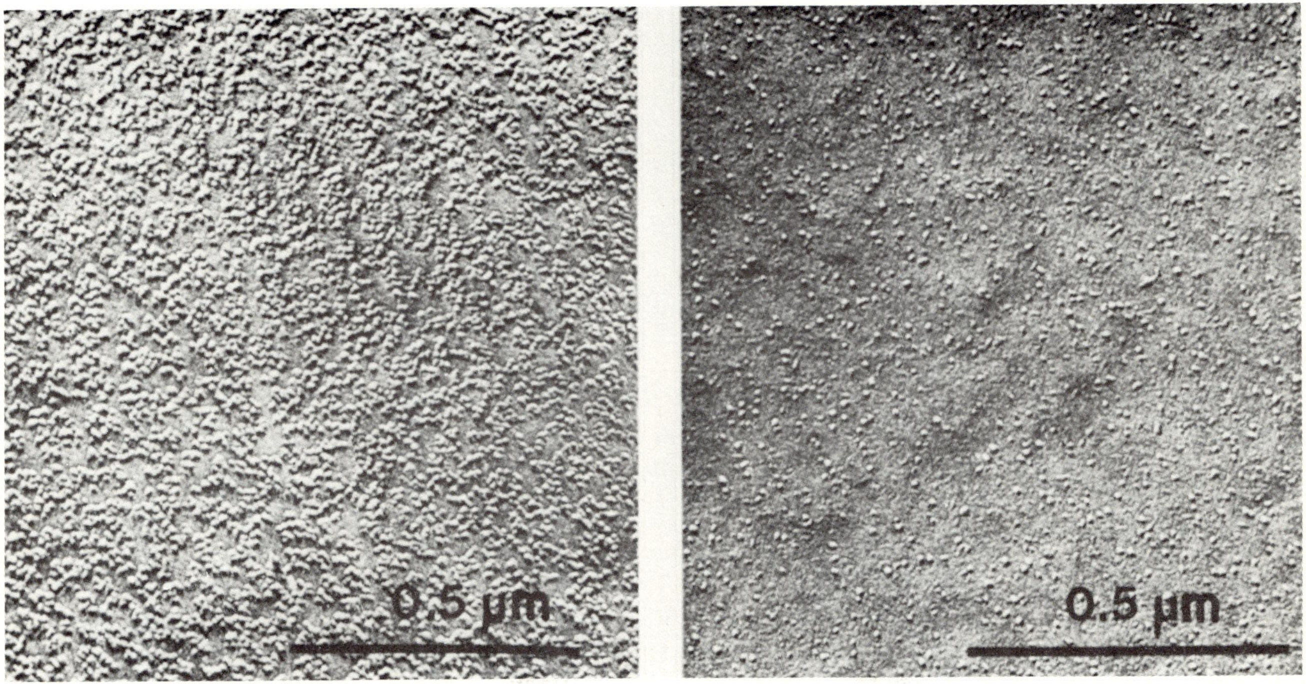

Figure 2-16 Freeze-fracture electron microscopic images of the erythrocyte membrane. (Left) Hydrophobic face of inner (adjacent to cytoplasm) monolayer. (Right) Hydrophobic face of outer monolayer. Bars represent 0.5 μm. The MAP are more numerous on the inner monolayer. (Reprinted with permission from *Prog. Biophys. Mol. Biol.,* vol. 26, N. S. McNutt and R. S. Weinstein, Membrane Ultrastructure at Mammalian Intercellular Junctions, Copyright 1973, Pergamon Press, Ltd.)

Another biologic tactic had been applied before this question about membrane protein location was recast in the context of the mosaic model: the use of enzymes to probe the cell surface. The enzyme neuraminidase had been shown to remove or inactivate the receptor for an influenza virus that attaches to erythrocytes. This receptor is the sialic acid group (*N*-acetylneuraminic acid in the human) of glycophorin—a major membrane glycoprotein mentioned in Section 2-3. Neuraminidase was shown to remove over 90% of the surface charge responsible for red cell migration in an electric field and a corresponding proportion of neuraminic acid. These observations confirm the notion that at least some of the carbohydrate portions of glycophorin are exposed at the outer surface, because neither virus nor neuraminidase penetrates the intact erythrocyte membrane.

Artificial reagents have also proved useful for this problem. A reactant that covalently binds to particular protein side chains or to the amino or carboxyl end of the peptide chain can be labeled (for example, with a radioactive atom) so that the protein with which it reacts can be located on an electrophoresis gel after separation from the membrane. Such experiments indicate which bands represent proteins exposed to the suspending medium insofar as the reagent does not penetrate the membrane of the intact cell. On the other hand, if the reagent does cross the membrane, it will mark susceptible proteins exposed at the cytoplasmic side as well. These experiments have produced strong evidence that different proteins are exposed at the inner and outer interfaces of the membrane, and many, if not all, of those exposed at the outer surface are also exposed at the inner surface.

The experimenter need not depend on membrane-permeable reagents. Erythrocyte membranes can be made permeable (p. 28) to ordinarily impermeable reagents and to antibodies and enzymes. The reactivity and labeling of proteins of ghosts not resealed (permeable) and intact red cells (impermeable) can be compared to determine whether a particular protein or group is accessible from inside, outside, or both. Such conclusions must be tentative, since hemolysis is known to change the accessibility of various chemical groups from the outside alone.

Another very clever extension of this methodology deserves mention here. Proteolytic enzymes known to hydrolyze peptide bonds adjacent to only specific side chains (that is, specific for particular amino acids) can be applied to the red cell membrane before or after labeling with specific reagents. This can be done with intact cells and with ghosts. The cells or ghosts can be washed free of enzyme, and the proteins and protein fragments can be extracted, separated, and identified to some extent. From these observations a detailed location and orientation of a protein and specific components of the protein can be deduced—at least with respect to exposure at inner or outer membrane surfaces and crossing of the membrane by portions of the peptide chains. Recent evidence of this type suggests that the peptide chain of band 3 protein crosses the membrane more than once.

Some major generalizations seem to be justified on the basis of the kinds of experiments described above. There are more distinguishable protein bodies exposed at the cytoplasmic face than at the extracellular face of the plasma membrane. All the carbohydrate groups of membrane glycoproteins are exposed to the face opposite the cytoplasm—the extracellular face for plasma membranes. Possibly all proteins exposed to the extracellular medium are glycoproteins or closely associated with glycoproteins. The proteins that span the membrane seem to have the amino end of the peptide chain at the side opposite the cytoplasm and the carboxyl end at the cytoplasmic side.

The total array of carbohydrate groups on the anticytoplasmic face of a biomembrane is often called the *glycocalyx*. It is responsible for important physical properties of cell surfaces and includes critical receptors, as will be seen in later chapters.

The sizes of the MAP seen by freeze-fracture electron microscopy are greater than can be accounted for by single proteins of molecular masses found by electrophoresis and ultracentrifugation. These particles are complexes of proteins. By treating extracted proteins in ways assumed to interfere specifically with hydrophobic bonding, hydrogen bonding, disulfide bonding, and so on, and subsequently performing electrophoresis, an experimenter can show corresponding decreases in higher-mass bands and increases in lower-mass bands. This leads to various conclusions about the associations of particular proteins. For example, glycophorin has been characterized by some investigators as a dimer in the membrane. Correlations between morphological observations of particle distribution and distributions of specific labels have led others to conclude that the MAP of the human red cell are complexes that include at least glycophorin and band 3 protein.

Caution is required in interpretation of these observations, for the components may react or dissociate very rapidly at almost any stage of the separations. The pertinent question for the biomembranologist is: Which proteins are associated *in the membrane*? To answer this question some investigators have treated the membrane with bivalent cross-linking reagents in order to form irreversible covalent bonds between protein monomers sufficiently close in the membrane. Such cross-linking must withstand subsequent treatments designed to prevent polymerization of proteins that were separate when in the membrane. Comparison of gels obtained in this way with those obtained with proteins extracted from untreated membrane gives some idea of native complexes. Some of these cross-linking reagents can be synthesized so as to have a known distance between two reactive groups. Application of these reagents allows estimates of distances between reactive groups on protein monomers when they were in the membrane.

The culmination of this rich variety of techniques used on the human erythrocyte membrane is illustrated in Figure 2-17. It is worth noting that the proteins of the membrane fall into three classes with respect to spatial relation to the lipid bilayer: (1) submembranous–the extrinsic, perhaps largely filamentous proteins on the cytoplasmic side of the bilayer; (2) half-penetrating–globular proteins interrupting the inner lipid monolayer, exposed to cytoplasm and/or submembranous protein at one pole; and (3) fully penetrating–globular proteins that span the bilayer, exposed to cytoplasm and/or submembranous protein at one pole and to extracellular space at the other pole.

The location of proteins in the hydrophobic region of the lipid bilayer raises questions about the nature of the surfaces of these protein molecules. The membrane is stable in an aqueous environment. Thus, the contiguity of protein with hydrophobic portions of amphiphilic lipids in the interior of the membrane and with hydrophilic portions and water at the boundaries of the membrane must represent a relative minimum of free energy for the system. This is reflected in the necessity for so-called chaotropic agents, such as detergents, which break up hydrophobic bonding, to extract integral proteins. Glycophorin of the human red cell membrane has been examined by amino acid sequence analysis. The results are consistent with an amphiphilic globular form; of the 131 amino acid residues found in one study, the first 72 at the amino end and the last 39 at the carboxyl end of the peptide chain are largely polar. The 20 in the middle portion form a hydrophobic region. This fully penetrating integral protein is therefore amphiphilic. A similar situation must exist with the half-penetrating integral proteins, except that one hydrophobic and one hydrophilic region must exist (as with the lipid components) rather than one hydrophobic region separating two hydrophilic regions.

This asymmetry in the distribution of membrane proteins in the lipid bilayer is not the only chemical asymmetry characteristic of red cell membranes. The distribution of phospholipids between the parallel laminae of the lipid bilayer is also asymmetric. Little can be said as yet about such an asymmetry in other biomembranes, but the situation in the human erythrocyte membrane is quite clear from labeling studies, side-specific enzyme digestion, and exchange studies. The choline phospholipids (sphingomyelin and choline phosphoglyceride) predominate on the outside and are absent or nearly so on the inside (cytoplasmic side). Ethanolamine phosphoglyceride and serine phosphoglyceride, both having free amino groups, predominate on the inside and are absent or nearly so on the outside.

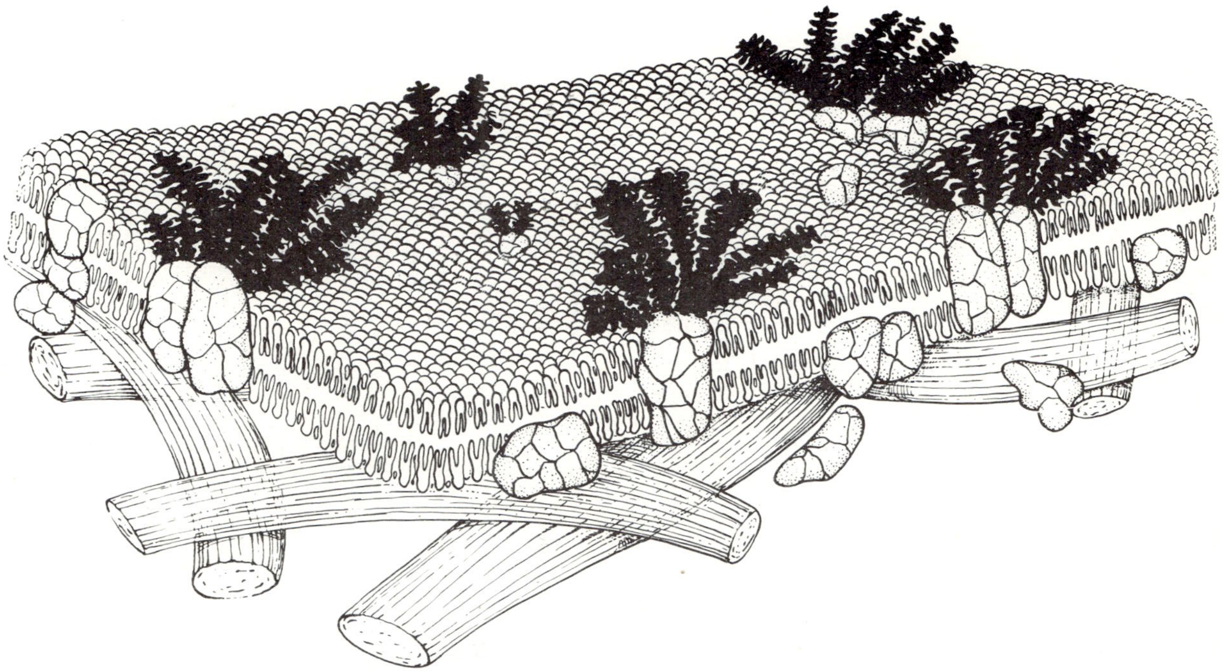

Figure 2-17 Molecular organization of the erythrocyte membranes. The lipid bilayer is shown as composed of the forked long phospholipid molecules and the somewhat smaller cholesterol molecules. Integral proteins are embedded in one or both lipid layers. Some of these are glycoproteins with the carbohydrate groups represented as projecting from the extracellular face (top). Peripheral proteins (mostly spectrin) are shown as predominantly filamentous at the cytoplasmic face (bottom).

Contradictory results exist, but the strongest evidence now supports the contention that phospholipids do not change position from one leaflet of the bilayer to the other (flip-flop) at detectable rates. There is some evidence suggestive of such a process for cholesterol and of asymmetry with respect to the amount of cholesterol in the inner and outer leaflets (more in the outer), but no firm conclusions can yet be reached.

In summary, then, the biomembrane is a laminar structure about 10 nm thick, stable in aqueous media, consisting of a phospholipid and cholesterol bilayer in which are embedded amphiphilic proteins (often as oligomers), and sometimes including a relatively weakly attached complement of filamentous protein at one face.

This model is the basis for the discussion of the physical properties of biomembranes in the next section and the various topics presented in the following chapters.

2-5 PHYSICAL PROPERTIES OF THE BIOMEMBRANE

Having established a more or less general molecular model of biomembranes, one can explore its implications in two directions. One direction is toward the specific components and their reactions with other specific entities, endogenous or exogenous—that is, toward molecular mechanisms and chemical reactions. These will be emphasized in later chapters in connection with specific biologic phenomena. The other direction is toward the behavior of biomembranes or closed sacs of biomembrane as materials and bodies—that is, toward behavior and physical interactions on a larger-than-molecular scale. These will be discussed briefly here.

The most obvious mechanical property of biomembranes is that they hang together under rather rough treatment. They can be bent, stretched, and compressed in an aqueous environment and still maintain their composition and properties as continuous laminar barriers. This is especially clear for the plasma membrane, which maintains the cell even as it is distorted by external forces. This characteristic of the plasma membrane has been described in the context of materials science. From this viewpoint the membrane is said to demonstrate viscoelastic behavior. It has elastic properties, since a deformation requires force in proportion to the extent of deformation, and the energy producing the deformation becomes potential energy of the deformed membrane and is returned to the environment when the deformation is allowed to vanish spontaneously. Classic examples are the moderately stretched spring and the moderately stretched rubber membrane.

The biomembrane also exhibits viscous behavior, wherein the force required to produce a given deformation is proportional to the rate of deformation, and the energy of the deformation is dissipated and not recoverable. A classic example is moving a rigid body through a fluid—for example, moving a hand through water. Stop the exertion, and the hand does not spontaneously return to its initial position. The water is also a little warmer. Since the biomembranes studied from this viewpoint display both sorts of behavior, they are said to be viscoelastic bodies.

Again, the biomembrane most thoroughly studied in this way is the red cell membrane. Its natural state as a single closed sac around a homogeneous liquid is particularly convenient. It can be deformed in controlled ways by micropipettes and by flows when attached to glass. In the pipette experiments the pressures used to deform can be observed and the forces calculated. In the flow deformation experiments the flows can be controlled and the deforming forces calculated. Knowledge of the molecular structure of the erythrocyte membrane is consistent with the observed viscoelastic behavior.

The hydrophobic attraction among membrane components can account for the maintenance of structural integrity under large deformations in aqueous media. These nonspecific attractions among the lipids of the bilayer and between lipids and proteins are an adequate basis for regarding the membrane as a two-dimensional liquid in which viscous behavior is a normal expectation. Elastic properties of such a structure are another matter. Two possible explanations for elastic behavior come to mind. One is that the components of the bilayer are more fixed relative to one another than the two-dimensional liquid model suggests. There may after all be a network of proteins or even of specific sequences of lipid molecules based on interactions that have not been detected. The other explanation seems more plausible and is supported by evidence cited in Section 2-4. The elastic behavior is consistent with the existence of a network of filamentous protein lying at the interface of membrane and cytoplasm. In the red cell there is the extrinsic protein that includes spectrin (Section 2-3). Such morphologically fuzzy layers have been observed in association with other plasma membranes.

Beyond such viscoelastic behavior, biomembranes do fragment and tear in the face of sufficient stress, either tension or shear (Figure 2-18). In the event of such a disruption of the continuity of the bilayer in an aqueous environment, the hydrophobic core of the membrane is exposed to water along an edge. This is an unstable situation; the free energy is increased because of the juxtaposition of water and hydrophobic regions of amphiphilic lipids and proteins at this edge. This unstable situation can be reversed by the resealing of a hole in a membrane or by the closing of an open membranous sac formed by a tear. The latter process has been observed in experimental situations in which micropipettes have been used

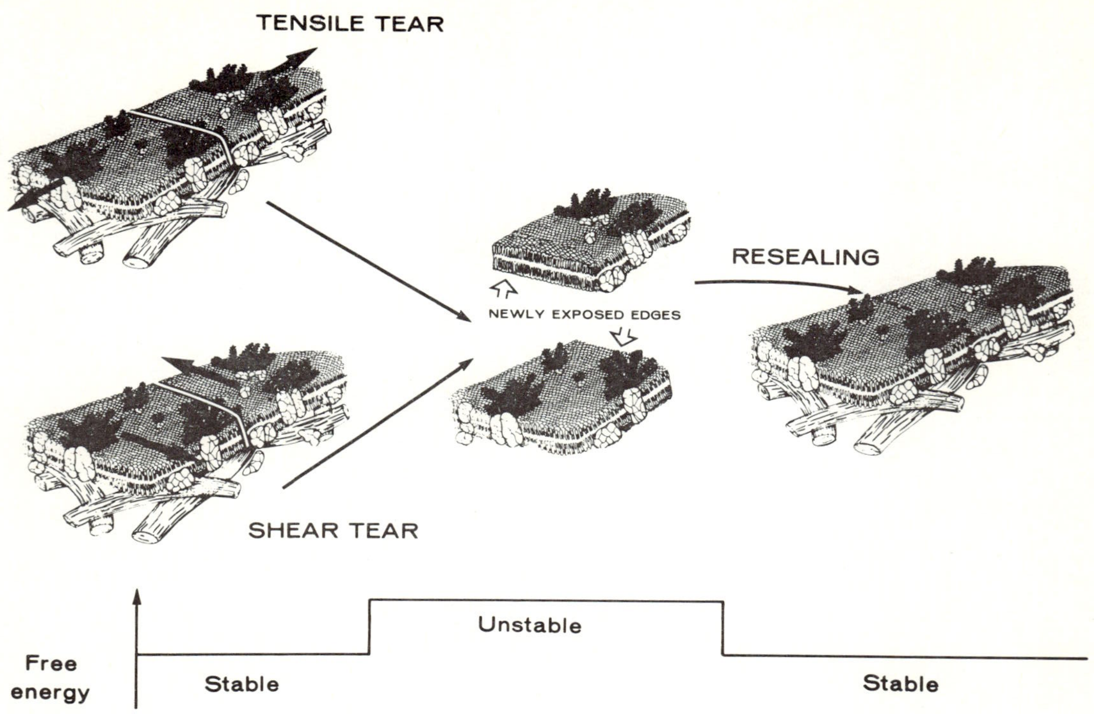

Figure 2-18 Mechanical disruption and resealing of a biomembrane in an aqueous medium. Shear and tensile forces (left) result in two pieces with edges exposed to water (center). Resealing restores the hydrophobic associations of the edges (right) and reduces the free energy represented by the respective heights of the line below the illustrations.

to pull off membrane that subsequently forms microvesicles by spontaneously resealing. Red cells caught on fibrin strands *in vivo* probably occasionally tear and subsequently reseal, forming two smaller cells. These are cases of reduction of free energy by reformation of continuous closed hydrophobic layers.

The electric properties of a biomembrane as a surface or body are critical to understanding cell behavior. These properties to some extent determine interactions between cells and between cells and noncellular surfaces. The properties of the liquid media adjacent to the membrane are also influenced by these electric properties. The plasma membrane will be the focus of this discussion since the electric properties of this membrane, exposed to the suspending medium, are most easily observed. Imagine a cell in isolation—an unreal but useful image. The system is electrically neutral, but the surface presents many chemical groups that ionize in water, especially the phospholipid head groups and the carbohydrate groups of glycoproteins. Now imagine the cell as it is immersed in water. Ionization occurs and charges appear. Some of these charges are the ionized groups of membrane components fixed in the membrane. This is called a *fixed charge.* The charge on the cell is matched by an equal total net charge in the medium contributed by ions freed from the cell surface. In the absence of any disturbance such as the translational and rotational kinetic energy of the molecules of the medium and membrane, these more or less free charges, *counterions,* would each be closely associated with fixed charges. For the most part one monovalent ion would be in the vicinity of a unit fixed charge of opposite sign. Electroneutrality would prevail on a small scale.

In general mammalian plasma membranes have largely negative fixed charges in aqueous media. Thus, the situation at very low temperatures would be well represented by a surface of largely negative fixed charges and just outside this a surface of largely positive counterions. At ordinary temperatures, however, the thermal motions of the system disperse the counterions so that the charge distribution differs significantly from the imaginary neat layer of fixed charge and juxtaposed layer of counterions. In an imaginary series of thin shells parallel to the surface of fixed charge, the density of counterions will be greatest in the closest shell and progressively less in more distant shells. Electroneutrality exists on a larger scale. The electric potential varies similarly: the potential is most positive in the closest shell and progressively less positive at greater distances from the surface of fixed charge. Figure 2-19 illustrates these effects.

Living systems are not this simple. The media surrounding living cells are aqueous solutions containing ions. These ions interact with the fixed charge of the cell surface and with the ions derived from the ionization of membrane components. In general, the more ions (that is, the denser the distribution of mobile positive and negative charges) in the medium

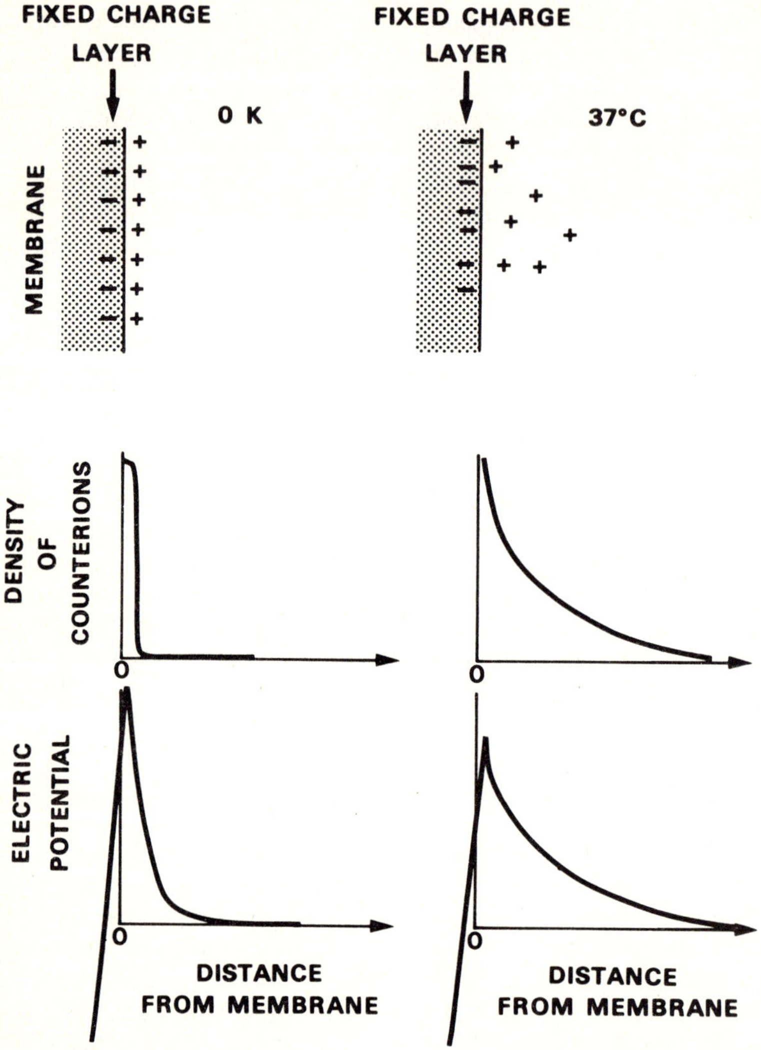

Figure 2-19 Electrostatic conditions in a medium bearing mobile positive charges adjacent to a surface bearing fixed negative charge. (Left) The hypothetical situation in the absence of thermal energy ($kT = 0$). (Right) The situation at physiological temperature.

the more compressed the variation in potential will be. At a given temperature the greater ionic strength of the suspending medium (more ions per unit volume) moves the curve describing the electric potentials (as in Figure 2-19) down and to the left. Usually, for mammalian cells at normal temperatures and in media of the usual ionic strengths, the potential is at $1/e$ (37%) of its maximum at a distance of a few tenths of a nanometer (about an ordinary small ionic radius) away from the surface of fixed charges.

The existence of these charges on and around cells causes them to react to imposed electric fields. Cells move through a potential gradient normally to the positive pole. The imposed field penetrates the cloud of counterions so that the cell reacts as a net negative body. The induced motion of the cell and the similarly induced opposite motion of positive counterions causes some of the counterions to be swept away, backward relative to the moving cell. The force resulting from the interaction of the charge on the cell and the field in the medium increases the cell's velocity until the viscous drag is an equal force in the opposite direction. This terminal velocity can be used to calculate the surface density of fixed charge given the temperature, ionic strength, and potential gradient in the observation chamber. This phenomenon is analogous to the electrophoresis of proteins described in Section 2-3. The measurement of terminal velocity is done by microscope, and the procedure is called microelectrophoresis. The surface charge is determined from the *zeta potential,* the electric potential at the surface of shear, which is calculated from the terminal velocity, potential gradient, ionic strength, and temperature.

The surface of shear is the surface outside the cell (biomembrane) that, on the average, describes the separation between the material that moves with the cell and the material that acts as part of the bulk medium. The fact that this surface of shear is not congruent with the cell surface is evidence of an important effect of the fixed charges and polar groups of a cell surface. Water molecules, being polar, interact with these fixed charges, the polar groups, and the counterions associated with them. There develops a layer of water at the membrane interface that is more ordered than the bulk water. The molecules are on the average more aligned in a specific way and relatively fixed. Depending on the distribution of such fixed polar groups and counterions and on the temperature and ionic strength of the medium, this layer of *vicinal* (in the vicinity of the membrane) water may be one to several molecules thick. This organized, relatively fixed layer is different from the bulk medium in significant ways. Transport of water and solutes may well be significantly limited; the concentration of H^+ may differ; the viscosity is probably greater; and the activities (probabilities of reaction) may differ for some solutes.

What is now known of the chemical organization of biomembranes, viscoelastic materials, and aqueous solutions gives us a fairly satisfactory basis for understanding much of observed cell behavior. There is a unity to the picture, together with a richness that promises to hold up well to the challenges of biologic problems.

2-6 SUMMARY

A biomembrane is an organized, three-dimensional array of molecules of phospholipids, cholesterol, and proteins that is 10 nm thick. The amphiphilic phospholipids and cholesterol are arranged in a bilayer with hydrocarbon chains contiguous in a hydrophobic interior and hydrophilic phosphoryl and hydroxyl groups exposed to water on opposite faces of the membrane. The proteins of the membrane (half or more of the membrane mass) are of several types: (1) filamentous and associated with the cytoplasmic face of plasma membranes; (2) globular with a central hydrophobic surface and polar ends and spanning the thickness of the membrane; and (3) globular with one hydrophobic surface region and one hydrophilic surface region and occupying a half-thickness of the membrane. Most, possibly all, the proteins that span the thickness of the membrane are glycoproteins with polar chains of sugars exposed to the medium opposite the cytoplasm. These sugars are collectively called the glycocalyx.

Some membrane components move readily in the plane of the membrane, although not across the membrane. The general nature of the array is one of flexibility and fluidity combined with organizational stability. The polar and ionizable groups at interfaces with water produce a significant surface charge and a layer of vicinal water.

Chapter 3

Transmembrane Transport

The report of Overton on membrane permeability in 1895, mentioned in Chapter 1, recognized the importance of the plasma membrane in determining the differences between cytoplasm and extracellular medium. This approach to the problem reflected a view of the plasma membrane as a passive barrier. But Overton's measurements of permeability did not entirely correspond with lipid solubility of the permeants. There were substances other than lipids that crossed the membrane with peculiar results. Sugars and cations, for example, are found at very different concentrations on opposite sides of biomembranes. Electrophysiological observations showed differences in electric potential across membranes and revealed orderly and reproducible changes in transmembrane potential that could not be explained by consideration of the biomembrane as a simple barrier.

In the 1950s two major lines of development became the dominant influences on concepts of biomembrane transport. One was the phenomenological approach along thermodynamic lines (continued from earlier kinetic attacks on the problem). The other was initiated by the discoveries of the relation between energy-yielding, enyzme-catalyzed reactions and transmembrane transport of cations, which led to the kinetic biochemical approach to the problem. A third (minority) view, in which the membrane played no role except to confine proteins, had formed earlier. It was based on developments in physical chemistry having to do with the binding of water molecules and ions by surfaces, gels, and macromolecules. This approach, the *sorption theory*, will be discussed in Section 3-3.

The major part of this chapter will deal with the explanation of transport phenomena in the context of the *membrane theory*, in which the membrane itself is assigned the functions leading to the asymmetries in distributions of soluble substances on the two sides of the membrane. This is in contrast to production of asymmetries by particular adsorption characteristics to the macromolecular cell contents.

The reasons transmembrane transport has received much special attention are clear. On the one hand the biomembrane and associated transmembrane concentration differences present a pervasive and fascinating problem. On the other hand biomembranes are a central feature of life. Living things are highly organized and strikingly different from their nonliving environments. When dormant, as a seed or spore, living things maintain this difference and organization by minimizing exchange of material with the environment. When active they live by means of controlled inflows and outflows of matter and energy. A substance may enter at one point and leave at another–for example, water in at the mouth and out by way of the lungs and kidney. Or the flow may include a change in form–for example, glucose with high energy in at the mouth and carbon dioxide with low energy out by way of the lungs. All such controlled exchanges are the results of either a nonspecific or a very selective barrier. The very selective barrier is always a biomembrane. Except for some special organisms that require other organisms in order to do what is normally associated with life (viruses, for example), all living things separate their interiors from their environments by at least one biomembrane.

Control of exchange with the environment is but a small part of the control imposed by biomembranes of complex organisms. Much of the control is among parts of the organism itself. The more complex a biologic system, the more adaptable and stable it becomes in a changing environment. This stabilizing complexity is achieved by increasing the numbers of, and decreasing energy differences between, metabolic steps, and by spatial compartmentation of components and temporal arrangement of processes. Compartmentation and separation of components is achieved by barriers to flow and by more or less specific transport of substances across biomembranes. The relations between transmembrane flows and distributions across membranes are discussed in Section 3-2 for a constant membrane, and the effects of changes in time of membrane properties on the transmembrane distributions are discussed in Section 3-3.

It should be remembered that the distributions considered here—the different concentrations on the two sides of a biomembrane—are themselves explanations of differences in, for example, some meter reading obtained when an aliquot of intracellular liquid is put into a machine as opposed to an aliquot of extracellular liquid. The methods of observation and the assumptions and theory behind the experiments are very much a part of the interpretation. This unavoidable interaction between scientist and subject is a large part of the dispute between sorption theorists and membrane theorists. It is very likely that both views have much to contribute.

3-1 STEADY-STATE DISTRIBUTIONS

Transmembrane Free Energy Change

The steady state is one in which the concentrations[1] at all points of the system do not vary in tme. It can be a very active state with many flows and reactions, but the amounts and distributions of each substance in any given volume are constant. This may result from inflows being equal to outflows, from chemical reactions connecting inflows and outflows of different chemical species, or from some combination of these means of maintaining constancy. However achieved, constancy of energy and material distribution within a region is steady state for that region. In this section we adopt the view that the observed steady-state distributions in and about biomembranes are real (as opposed to faults in measurements) and are consequences of the properties of the membranes. Thus, in the simplest case of a single biomembrane we assume that the concentrations of dissolved substances are uniform throughout the continuous aqueous phase on either side of the membrane. The problem of explaining or predicting differences in concentrations on the two sides of the membrane then is one of interplay between the physical forces responsible for transmembrane flows and the relevant properties of the membrane.

From the viewpoint of the discussion of stability in Section 2-1, the movements of molecules and ions are determined by their relative free

[1] In the discussions of this section, the variable *concentration* is substituted for the thermodynamic variable *activity*, which is directly proportional to concentration. This introduces no conceptual inaccuracies. Furthermore, under conditions predominant in human tissues, concentration is very nearly equal to activity. If activity is thought of as a variable that expresses the probability of a particular species undergoing a transition from one state to another, then concentration is clearly one factor in activity. The more potential participants in an allowed transition, the more probable the occurrence of that transition in a unit of time.

energies[2] in the two locations (or environments) between which they move. A molecule will move from side 1 of a membrane to side 2 if its free energy is less on side 2, provided the molecule is not required to enter an intermediate state at a free energy higher than the initial state by much more than the thermal energy (Figure 3-1). The thermal energy is the mean translational and rotational energy of the component molecules and is reflected by the temperature of the system. For a given molecule or ion this thermal energy will fluctuate about the mean of all components of the system, which are in thermal contact. If two molecular states differ in free energy by not much more than this thermal energy, then the molecule of the type concerned will move back and forth between these states. The time-average distribution of the aggregate of all molecules and ions among states may be represented as in Figure 3-2 for a system such as that characterized in Figure 3-1.

[2] The thermodynamic state variable applied here is Gibbs free energy. See Annotated Bibliography, Thermodynamics, for references.

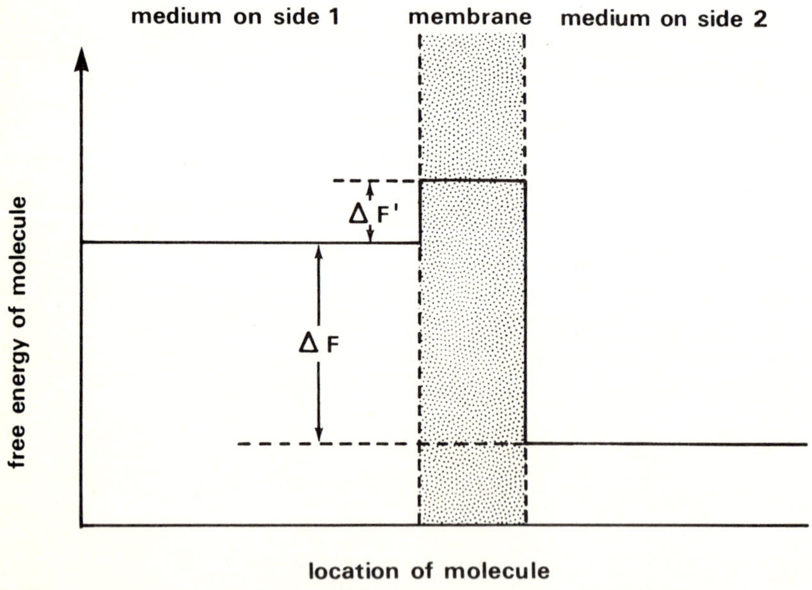

Figure 3-1 Free energy of a molecule in and around a membrane. There are three values associated with the three locations: side 1, within the membrane; and side 2. ΔF is the decrease in free energy of a molecule moving from side 1 to side 2. $\Delta F'$ is the free energy required to enter the membrane from side 1 (analogous to the activation energy of a chemical reaction).

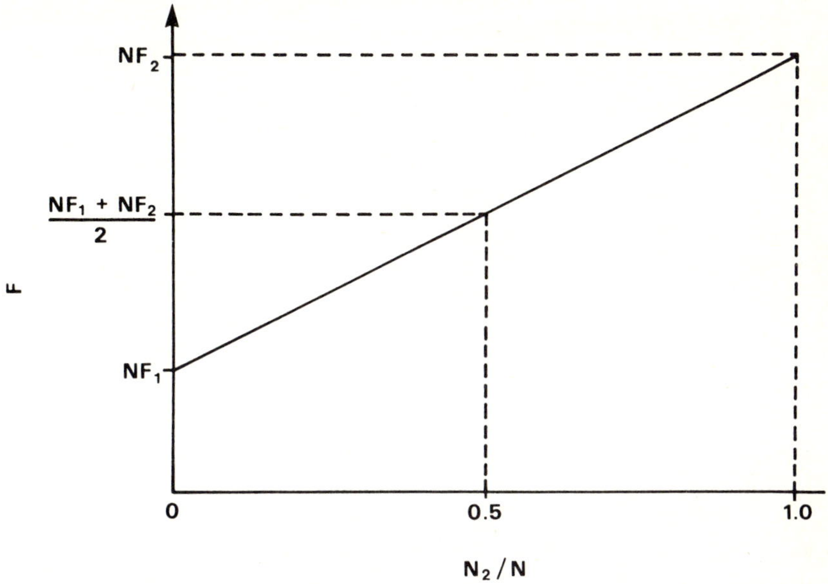

Figure 3-2 Free energy of a system of N identical molecules in and around a permeable membrane. The number of molecules within the membrane is negligible. $N = N_1 + N_2$. Total free energy $F = N_1F_1 + N_2F_2$, where F_1 and F_2 are free energies on the two sides and N_1 and N_2 are the numbers of molecules on the two sides. $F_2 - F_1$ is ΔF of Figure 3-1.

The free energy of a system is a minimum at equilibrium. For a steady-state system not at equilibrium there will be a higher time-average free energy determined by the energy flow into and out of the system.[3] Nevertheless, we may look at an individual molecule or ion as a system in itself and consider its movements as a process in which a system able to lose free energy to its environment alters its state (location) to minimize its free energy. This enables us to conveniently characterize the forces governing the distribution of a particular species of molecule or ion.

The influence of temperature has been noted with respect to the whole system (membrane plus adjacent aqueous phases). Temperature, as manifest in the kinetic energy of a molecule, also influences the free energy of a particular molecule. For organisms at constant environmental temperature or cells within homothermic animals, however, one need not consider

[3] Such a system will be in a state of minimum entropy production provided it is not too far from equilibrium, according to the premises of the theory of thermodynamics of irreversible processes. See Annotated Bibliography, Thermodynamics, for appropriate references.

Table 3-1 Pertinent Driving Forces

Driving force	Variable	Comment
Concentration gradient (chemical potential gradient)	Concentration (activity)	Concentration determines the probability of substrate interaction or transport related to presence of substrate at a location.
Electric potential gradient	Electric potential	The associated change in energy results from attraction and repulsion of electric charge. This is associated with current (flow of charge).
Pressure gradient	Pressure	The associated change in energy is caused by mechanical force coupled with volume change—most often observed as bulk flow.

temperature except to say that it is assumed to be uniform throughout the system considered. The other characteristics of the system that influence the free energy of a molecule or ion at a point are concentration, pressure, and electric potential. Others are generally of no consequence in biologic systems. Table 3-1 is a summary of these *driving forces*, including their chief characteristics to be used here for explaining transmembrane transport. The notation used in giving the relations discussed in the following sections is chosen to be consistent with best current usage in biophysical and physiological writing. The reader may wish to devise an alternative notation. In any case it is the quantitative relations among physical variables and parameters that are important.

Diffusion

Diffusion is the movement of a molecule or ion through the medium in which it is dissolved *relative to* the surrounding components of the medium. It differs from movement *with* the surrounding medium, which is bulk flow. Diffusion produces a flow that is proportional to concentration gradient. The proportionality constant is the *coefficient of diffusion.* Thus we have Fick's law of diffusion:

$$J_s = -D \frac{dc_s}{dx} \tag{3-1}$$

where J_s is the flux (amount per unit area per unit time) of solute, D is the diffusion coefficient (unit length squared per unit time), and the gradient dc_s/dx is the derivative of concentration with respect to distance (amount per unit volume per unit length). For clarity, consider gradient, flow, and position in only one dimension. The gradient is defined so that it is positive when concentration c increases with increasing distance x. If the gradient is constant (concentration varying linearly with distance) over a distance $\Delta x = x_2 - x_1$, then $dc_s/dx = \Delta c_s/\Delta x$ where $\Delta c_s = c_2 - c_1$, and c_2 and c_1 are the concentrations at x_2 and x_1, respectively (Figure 3-3).

Often in the situation wherein a biomembrane separates homogeneous solutions, the concentration difference Δc is taken to be that between the two aqueous media and the distance Δx is the thickness of the membrane. If the rate-limiting step of transport is diffusion through the membrane (well-mixed media on both sides), and the partition[4] of solute between

[4] The partition of a solute in two adjacent media is given quantitatively by the ratio of the concentrations of the solute in the two media at equilibrium at the given temperature.

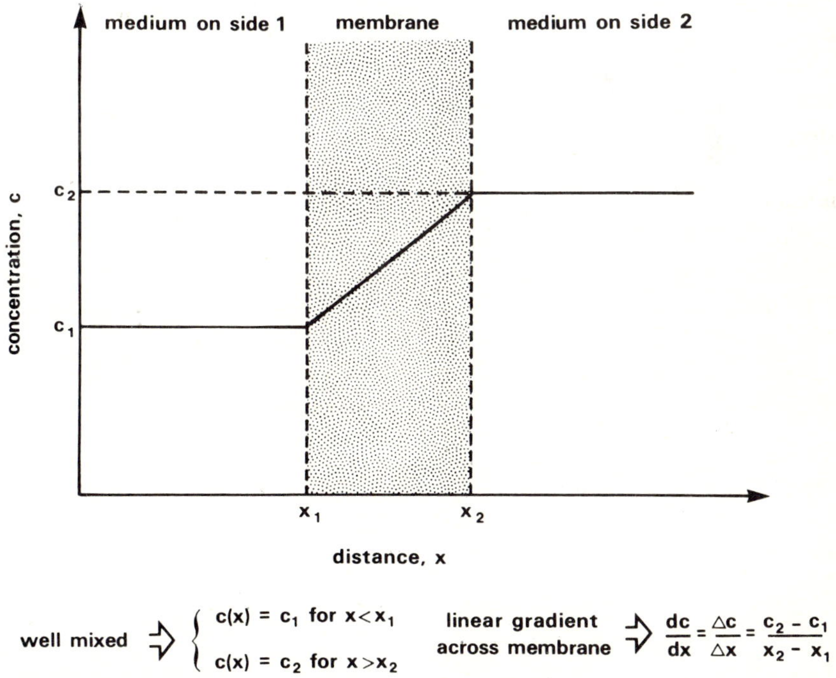

Figure 3-3 Notation for representing the transmembrane concentration gradient.

media and membrane is incorporated in the constant D', then the flux is

$$J_s = -\frac{D'}{\Delta x} \Delta c_s \tag{3-2}$$

A frequently used convention is to define the *permeability coefficient* for the particular molecule and the particular membrane as $P = D'/\Delta x$, whereupon

$$J_s = -P \Delta c_s \tag{3-3}$$

where P has dimension unit length per unit time. (The units of P are usually cm/s, in which case J_s is in mol/s·cm^2.)

Bulk Flow, Filtration, and Osmosis

A pure, permeative solvent passes through a membrane with a flux proportional to the pressure difference across the membrane. In physiological systems the volume flux is essentially equal to the water (solvent) flux, and

$$J_v = L'_p \bar{V}_w \Delta P \tag{3-4}$$

where J_v is volume flux (volume per unit time per unit area), $\bar{V}_w$ is the partial molar volume of water (unit volume per mole), and ΔP is the pressure difference across the membrane (force per unit area). The sign of ΔP is assigned as with Δc (compare Figure 3-3). The positive constant of proportionality L'_p is defined by $L'_p \equiv L_p/RT$, where R is the gas constant, T the absolute temperature, and L_p the *hydraulic water permeability* or *filtration coefficient* with dimension of distance per time (usually cm/s). The convenience of so defining L'_p will be seen below.

Now consider a solution and a membrane very permeable to the solvent. The behavior of the system depends on the permeability of the membrane to the solute. If the membrane is as permeable to solute as it is to solvent, then bulk flow of solvent (volume flow) is associated with a solute flow:

$$J_s = J_v c_s \tag{3-5}$$

This J_s is distinct from diffusive J_s of Equation (3-3).

At the other extreme, the membrane may be totally impermeable to a solute. In such a case volume flow through the membrane produces a concentration difference across the membrane. For example, let the system initially be composed of a membrane permeable to solvent and impermeable to solute separating identical solutions at equal concentrations. Then an added pressure is applied on one side; solvent flows through the membrane and solute is retained on the upstream side. On the high-pressure side the concentration *of solvent* is lowered, and on the low-pressure side it is raised. This concentration gradient drives a flow of solvent (by diffusion) in the opposite direction. Conversely, the initial condition may have been a solute concentration difference. This implies a solvent concentration difference that drives a solvent flow across the membrane. This latter process is called *osmosis,* and the pressure difference that just balances the osmotic flow—so that the opposite flows driven by ΔP and by Δc_s are equal—is called the *osmotic pressure.* Osmosis, where a pressure is produced by a hydrostatic head or by tension in a closed membrane, is a dynamic relation between a diffusive flow, produced by a concentration gradient, and a bulk flow, produced by a pressure gradient. Where the flow is not captured in a gravitational field or closed cavity, no pressure gradient arises to produce a back flow and osmosis continues until the solvent concentrations are equal.

Although the osmotic pressure may be more clearly seen as a consequence of solvent concentration gradient (difference in solvent chemical potential or free energy), it is usually expressed as a function of total solute concentration difference Δc across the membrane, as in the theoretically derivable and experimentally verifiable van't Hoff equation:

$$\Delta\pi = RT\,\Delta c \tag{3-6}$$

The osmotic flow may be thought of as a pressure-driven flow as in Equation (3-4). The virtual gradient $\Delta\pi$ plays the role of ΔP, but the flow is in the direction of *increasing* solute concentration, so from Equations (3-4) and (3-6):

$$J_v = L'_p \bar{V}_w\, RT\,\Delta c = L_p \bar{V}_w\,\Delta c \tag{3-7}$$

For many important solutes and most biomembranes, neither extreme holds—neither perfect semipermeability nor entirely free passage of solute. Solute passes through the membrane but is partially filtered out of the passing solution. To account for such intermediate behavior, and to describe the extremes as well, a parameter σ called the *reflection coefficient* is defined. For special conditions it may be represented by

$$\sigma = \frac{\Delta\pi_{\text{obs}}}{\Delta\pi_{\text{th}}} \tag{3-8}$$

where $\Delta\pi_{\text{obs}}$ is the observed osmotic pressure difference when $J_v = 0$ and $\Delta\pi_{\text{th}}$ is the theoretical osmotic pressure difference calculated from the van't Hoff relation [Equation (3-6)]. The reflection coefficient is dimensionless and varies from 0 to 1 (there are exceptions) depending on the membrane and solute. By using σ, volume flux caused by osmosis is

$$J_v = L_p' \bar{V}_w \, \Delta\pi_{\text{obs}} = L_p' \bar{V}_w \sigma \, \Delta\pi_{\text{th}} = L_p' \bar{V}_w \sigma RT \, \Delta c = \sigma L_p \bar{V}_w \, \Delta c \tag{3-9}$$

The bulk flow contribution to solute flow is similarly corrected:

$$J_s = (1 - \sigma) c_s' J_v \tag{3-10}$$

where $1 - \sigma$ is the fraction not reflected. The concentration factor c_s' is an effective concentration dependent on how the concentration varies within the membrane. It differs, but often very little, from c_s on the upstream side of the membrane. This last modification is not established by any argument presented here, but it can be demonstrated on other grounds (see Annotated Bibliography). Similarly, the addition of the individual contributions to J_v and J_s is justified in another context. To sum up the major factors governing flux of a single ($\Delta c = \Delta c_s$) nonelectrolyte and water across a given biomembrane [from Equations (3-3), (3-4), (3-9), and (3-10)]:

$$\begin{aligned} J_s &= (1 - \sigma) c_s' J_v - P \, \Delta c_s \\ J_v &= -L_p \bar{V}_w \left(\frac{\Delta P}{RT} - \sigma \, \Delta c_s \right) \end{aligned} \tag{3-11}$$

The parameters σ and P are characteristic of the particular membrane-solute combination. The parameters characterizing the system are described in Table 3-2.

A Practical Corollary

In plants, where turgid cells may produce 10 atmospheres in pressure, the pressure gradient is ordinarily a very important factor in the distribution of water and solutes. But in animal tissues the pressure difference across biomembranes is normally very small. The above relations of solvent and solute flows to the pressure gradient are nevertheless important in under-

Table 3-2 Parameters Characterizing a Membrane-Solution System

Symbol	Name	Dimension	Unit in cgs system
P	Permeability coefficient	Length/time	cm/s
σ	Reflection coefficient	—	—
L_p	Hydraulic water permeability or filtration coefficient	Length/time	cm/s

standing the significance of certain transport-related specializations of animal biomembranes discussed below and in emphasizing certain restraints in such medical procedures as intravenous fluid therapy. For example, it is clear that intravenous replacement of water deficit in a patient must be done with an isotonic[5] solution such as glucose in water—not with pure water. Pure water would produce a water concentration gradient across the erythrocyte membrane in the region of the infusion; the cells would swell and become spherical; the membranes would experience a tension because of the pressure gradient induced; and the membranes would rupture or become permeable to hemoglobin and all other contained solutes (hemolysis).

This is clear enough, but it has been disastrously ignored that some solutes used in intravenous infusion (such as ethanol in the treatment of methanol poisoning) are characterized by transmembrane permeability coefficients approximately equal to or even greater than that of water. For example, infusion of an isomolar solution of ethanol induces a large concentration gradient for ethanol across the erythrocyte membrane ($\Delta c \gg 0$). The ethanol, being highly permeative ($P \gg 0$) for the membrane, rapidly crosses the membrane. In addition, the increased transmembrane gradient in water concentration contributes to swelling of the cell. Hemolysis occurs with harmful consequences.

[5] Tonicity and osmolarity are different. Osmolarity is a measure of total solute concentration and characterizes a solution per se. Tonicity is meaningful only when referring to a specified membrane (cell) as well as a solution. By definition, an isotonic solution induces no volume flow across the membrane; a hypotonic solution produces volume flow *from* the solution across the membrane; a hypertonic solution induces volume flow *into* the solution across the membrane. In clinical medicine the erythrocyte is the standard for tonicity (because of intravenous fluid therapy). A hypotonic solution swells erythrocytes; a hypertonic solution shrinks erythrocytes.

Transmembrane Electric Potential

The influence of electric potential (energy per unit positive charge resulting from electromagnetic forces) on ion transport remains to be considered. Or, to look at the inverse form of the problem, how transmembrane transport of ions influences transmembrane electric potential must be considered. The electrostatic force tends to separate like charges and to aggregate unlike charges. Therefore the potential energy (and thus the free energy) of a positive charge is greater in a region of relatively greater positive (lesser negative) charge than in a region of lesser positive (greater negative) charge. The influences of concentration gradient, pressure gradient, and electric potential gradient are additive. So it is possible in principle to have steady states in which concentration differences are balanced by a potential gradient, in which a potential difference is opposed by a concentration or by a pressure difference or by both, or (as seen above) in which a concentration gradient is balanced by a pressure gradient (osmosis).

The influence of an electric potential difference across a membrane on transmembrane transport of ions can be modeled after Ohm's law[6]:

$$\Delta\psi = iR \tag{3-12}$$

where $\Delta\psi$ is the potential difference between two points in a conductor, i is the current (flow of charge) between the two points, and R is the resistance between the points, a property of the conductor. If the conductor is a biomembrane separating two solutions and R is the resistance of a unit area of membrane, the current carried by a flow of ions across a unit area will be

$$i = zFJ_s = -\frac{\Delta\psi}{R} \tag{3-13}$$

where z is the ionic charge, F is the Faraday constant (charge per mole), J_s is the ion flux (mole per unit time per unit area), and i is taken positive in the direction of decreasing ψ ($\Delta\psi < 0$, $\Delta x > 0$). The resistance is assumed to be entirely in the membrane and the potential to be equal throughout the volume on a given side. The flow of a given ion is directly proportional to the electric potential difference and inversely proportional to the resistance of the membrane. The ion flux J_s in Equation (3-13) is shown to be dependent on $\Delta\psi$ for a given membrane. J_s is also dependent

[6] Ohm's law is a special case of coupled flow and force in the spirit of the thermodynamics of irreversible processes. It is valid sufficiently near equilibrium (at which i and $\Delta\psi$ vanish).

on ΔP and Δc (unless $L_p = P = 0$, in which case J_s and i would both be zero except for active transport described in the next section). For very large values of R, i can be negligibly small while $\Delta\psi$ is appreciable.

In mammalian systems cells generally exhibit nonzero potential differences across plasma membranes. The cells most studied with respect to electric properties are nerve cells and muscle cells. More will be said of these in Section 3-3. For now it should be noted that the description of cells and cellular activity in terms and symbols taken from the theory of electric circuits implies nothing special about the cells or membranes. Current corresponds to ion flow; conductance (inverse of resistance) corresponds to permeability; and electric potential difference corresponds to concentration gradient (distribution of ions). The usual electric description (of nerve, for example) does, however, include transient changes that reflect variation in certain parameters introduced above. Such changes are likely to compromise the validity of the equations given here.

Component-reactive Transport

Another feature of transmembrane transport in the steady-state condition is a peculiarly biologic one. In plasma membranes, membranes of the nuclear envelope, mitochondrial membranes of animals, chloroplast membranes of plants, and membranes of single-celled species, substrates have been observed to cross the membrane by a mechanism that appears to include the interaction of the substrate with a specific constituent of the membrane. The membrane constituent involved may be a permanent (constitutive) component of constant concentration or an adaptive component that increases in concentration within the membrane some definite time after the cell is exposed to the specific substrate.

The substrate-component interaction is one of specific bonding. Some of these cases are associated with transport from a lower (on one side of the membrane) to a higher (on the other side) free energy state (lower to higher concentration, lower to higher electric potential, or the like). Such endergonic substrate translations are associated with exergonic metabolic reactions. All such events, whether energy-requiring or not, involve the following steps, which are distinguishable in principle and sometimes experimentally: (1) recognition-binding, (2) transmembrane translation, (3) release of substrate, and (4) resetting of transport mechanism. These steps are represented in Figure 3-4 for a hypothetical carrier mechanism. The energy input in the endergonic cases may in principle occur at any of these steps. For the sake of brevity we may refer to this as *component-reactive transport.*

Figure 3-5 shows the free energies of substrates at various stages of

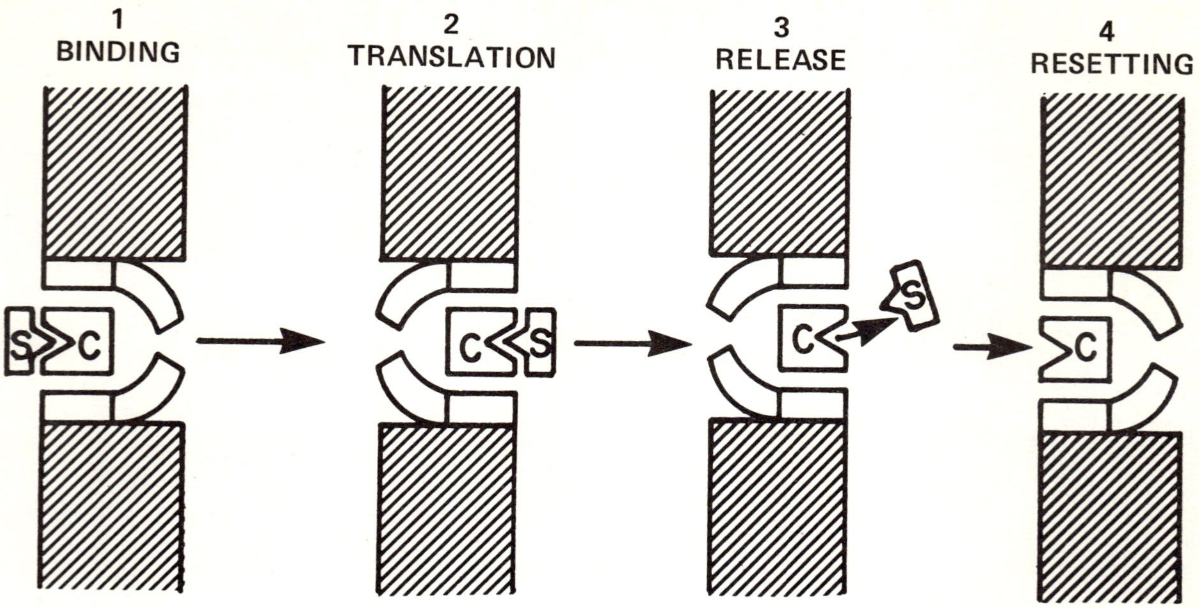

Figure 3-4 Steps in component-reactive transport. The mechanism shown is a hypothetical one that might be accommodated by conformational changes in integral proteins. S, substrate; C, carrier component of membrane.

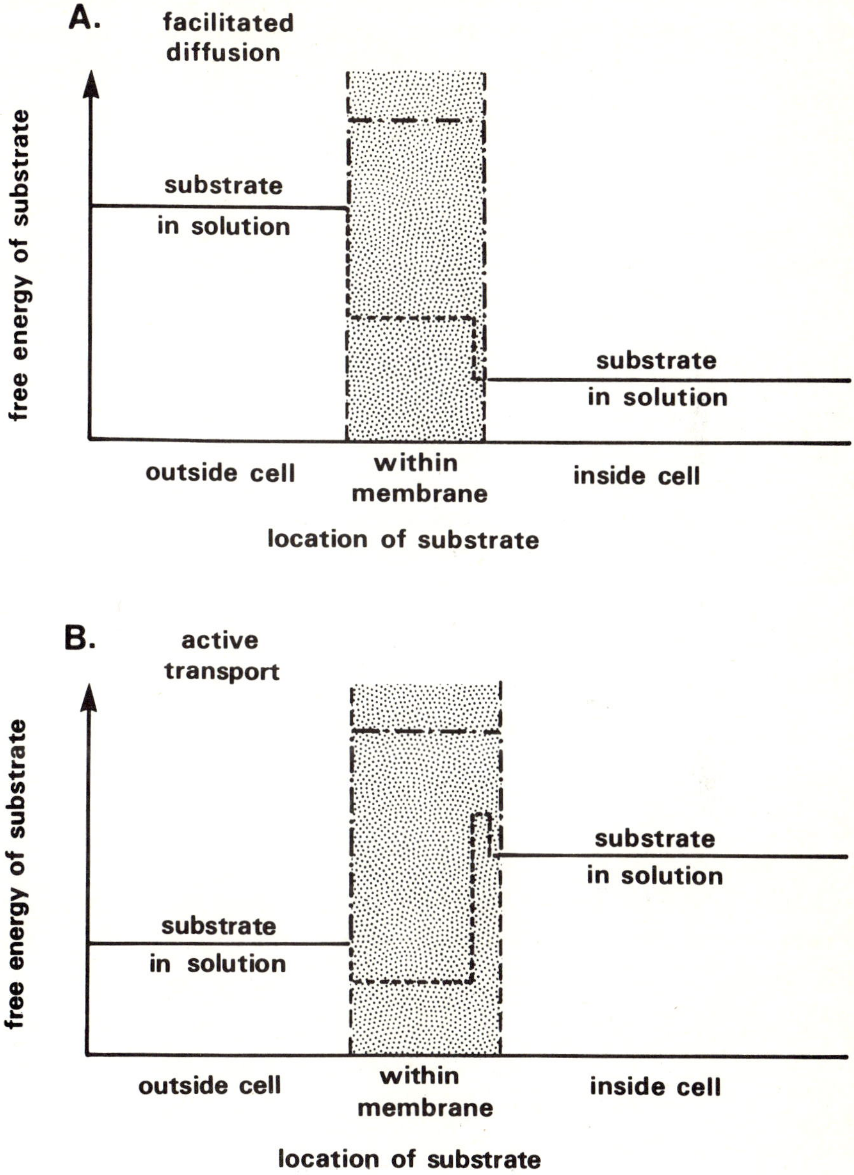

Figure 3-5 Energetics of component-reactive transport. (A) Facilitated diffusion. (B) Active transport. Alternative possible energy-paths are shown through the membranes. The free energies within the membrane should be the free energy of the substrate-component complex minus the free energy of the uncomplexed component (carrier).

component-reactive transport. The energetics of this process form a useful basis for classification. If the result is the movement of a nonelectrolyte from a region of higher concentration to a region of lower concentration on the two sides of the biomembrane, then the energy of transport can be derived from the translation itself—provided the substrate-component reaction is not restrictively endergonic or provided the reaction energy is sufficiently recovered on release or resetting. If the transport steps are all exergonic, or strongly coupled and overall exergonic, or if the steps require energies approximating thermal energies, then the process is driven by the concentration gradient. This subclass of component-reactive transport is conventionally referred to as *facilitated diffusion*, and the membrane component involved is often called a *carrier*.

The essential kinetic nature of facilitated diffusion is that the flux of solute is directly proportional to the concentration difference across the membrane (as in *simple diffusion*) in the direction from higher to lower concentration. In some situations (for example, glucose and human erythrocyte) this provides a second and faster diffusion pathway, so that

$$J_s = -(P + P_f)\,\Delta c_s$$

where J_s is flux by diffusion, P is the permeability coefficient, and P_f is the coefficient of proportionality for facilitated diffusion. This equation is based on the assumption that the carrier is uniformly distributed over the membrane so that the number of carriers is proportional to the area. This can be a very significant cellular adaptation where the substrate is strongly partitioned at the aqueous medium-membrane interface in favor of the medium, thus inhibiting simple diffusion at the high-concentration interface. An experimentally critical distinction between facilitated and simple diffusion is that the former can be saturated—that is, the transport process reaches a maximum beyond which an increase in gradient is not accompanied by an increase in transport.

The other class of component-reactive transport is that in which the membrane is capable of moving a substrate from a lower to a higher free energy state. Such systems are coupled to energy-yielding metabolic reactions. The most studied and apparently widespread energy-yielding reaction is the hydrolysis of adenosine triphosphate (ATP) to adenosine diphosphate (ADP). There are several, perhaps many, distinguishable ATPases associated with transmembrane transport. Two prominent ATPases in animals, requiring Mg^{2+} as a cofactor, are involved in the transport of Ca^{2+} out of mammalian cells in one case and the coupled transport of K^+ into and Na^+ out of cells in the other. The latter seems to be a one-to-one exchange in many nerve cells, while in mammalian red blood cells three Na^+ are exchanged for two K^+. The K^+ receptor is at the outer face of the membrane; the ATP-hydrolyzing enzyme seems to be at the inner face. Other inorganic substrates transported against gradients in this manner apparently include NaCl across the mammalian gall bladder epithelium and Cl^- in the mammalian nephron.

The so-called Na^+-K^+-ATPase system has been isolated from membranes and the component reactions investigated. It seems to be a complex process involving separable proteins binding Na^+, K^+, and ATP at distinguishable stages. Interestingly, if larger than normal gradients in the normal direction are artificially produced across certain biomembranes, reverse (with respect to normal operation of the mechanism) flow of Na^+ and K^+ occurs. This is accompanied by the generation of ATP from ADP. The system is reversible *in toto.*

This latter type of component-reactive transport is often called *active transport.* This term should be used with caution because it has different meanings among writers of books and journal articles. The term is more specific if it is reserved for a normally energy-requiring transport (that is, against a free energy gradient) that is associated with substrate reaction with a normal component of the biomembrane. Some authors simply require the energy-requiring criterion. The next subsection describes a situation in which the energy requirement without reaction with a membrane component appears.

The transport of a substrate dependent on a metabolic reaction such as the transport ATPase systems does not fit into the equations expressed only in terms of the physical driving forces. To account for this type of transport coupled to a chemical reaction, one may introduce a term in the first of Equations (3-11) expressing the portion of J_s that is proportional to the coupled reaction rate. This rate is often expressed as a flow J_r—a flow through the specific reaction sequence.

In more informal terminology, these active transport mechanisms associated with ATPases are called "pumps," and the simple diffusion processes associated with permeability coefficients are called "leaks." For actively transported solutes the concentrations on the two sides of biomembranes are the result of the interplay of pumps and leaks. For example, the typical steady-state electric potential difference of several millivolts across a mammalian plasma membrane may be explained readily even if the pumping of Na^+ out and K^+ in is one-to-one. The permeability coefficient for K^+ diffusion across the membrane is greater than that for Na^+. The K^+ leak is driven inward by $\Delta\psi$ and outward by Δc; the Na^+ leak is driven inward by both $\Delta\psi$ and Δc. However, the higher permeability for K^+ and lower permeability for Na^+ makes it possible for the same one-to-one (or three-to-two) Na^+-K^+ exchange pump to maintain the observed gradients. The active transports of K^+ and Na^+ are coupled in the pump. The K^+ and Na^+ leaks are coupled in the common driving force $\Delta\psi$. The concentrations at steady state differ because the permeability coefficients P differ for the two cations.

Flow-Coupled Transport

This last category of transport may be descriptively referred to as *flow-coupled transport.* This, too, may occur so as to either increase or decrease the free energy of the transported substrate. The physiochemical events coupling flows referred to here may occur (1) within the membrane or (2) in the medium on one or both sides of the membrane.

When coupling occurs in the membrane, a substance interacts with another substance moved directly by one of the driving forces so that it is dragged along through the membrane. One suggested phenomenon of this class is *solvent drag,* whereby a solute molecule is carried via an aqueous channel (pore) through the membrane by the flow of water. In this case the driving force for the water is ΔP or $\Delta\pi$ (Δc), directly. The solute flow is coupled to J_v and is only indirectly driven by ΔP or $\Delta\pi$.

Whether a solute flow is considered flow-coupled or driven by one or more of the driving forces (ΔP, $\Delta\pi$, or $\Delta\psi$) indirectly through some other solute flow is often a matter of viewpoint. For example, J_{Cl^-} may be thought of as a flux in which Cl^- follows actively transported Na^+ by electrostatic attraction or as J_{Cl^-} driven by $\Delta\psi$ produced by active cation transport. The former view may be more useful to an enzymologist studying the Na^+-K^+ pump. The latter may be more useful to the biophysicist concerned with transmembrane electrokinetics.

The case where the coupling events occur outside the membrane may be considered as not a membrane phenomenon at all. However, the effect is just like the former case. The usefulness of the concept becomes clearer in applications to complex systems composed of differing membranes (Chapter 4). Consider a membrane impermeable to solute Y with a mechanism for transporting some compound XY in response to Δc_{XY}. If on one side of the membrane conditions push the reaction $X + Y \rightleftharpoons XY$ toward formation of XY and on the other side conditions (for example, removal of X) push the reaction toward dissociation, then there will be a net transport J_Y coupled to J_{XY} driven by Δc_{XY}.

The kinetic characteristic of flow-coupled transport of a substrate is that it is proportional to a flow of another substrate, whatever the mechanism of that flow may be. So the flow of another substrate, J_i may be added to $\Delta\psi$, ΔP, Δc, and J_v on the list of driving forces for J_s.

Whether certain sugars and amino acids taken into various mammalian cells against concentration gradients are examples of active or flow-coupled transport is a long-standing controversy. Greater concentrations of these metabolites are found inside than outside of many types of cell.

Summary of Steady-State Conditions

This section on steady-state distributions will benefit from a summing-up and a classification of transport mechanisms.

We may visualize the balance among transport processes that characterize a living mammalian cell as follows. The metabolism of the cells supplies metabolic energy sources, including ATP. The cell is provided with substrates such as glucose and amino acids at higher concentrations in the cytoplasm than in the extracellular medium. The endergonic transport of these substances is either flow-coupled or active. The ATP is hydrolyzed at sites inside the plasma membrane, thus supplying energy for the operation of the Na^+-K^+ pump and for the Ca^{2+} pump. These cation pumps operate against concentration gradients, and Na^+ and Ca^{2+} are transported against an electric potential gradient. The pumps operate continuously and balance (in the steady state) the corresponding leaks (that is, diffusion down the gradient). Other pumps may operate, some only in particularly specialized cells. The controlling enzymes are provided by the protein-synthesizing process in the cytoplasm regulated by the genetic apparatus of the nucleus and by transport across the nuclear envelope and its inner and outer membranes. The result is a metabolizing cell with total solute concentration greater inside than in the surrounding medium. The cell experiences a negligible pressure gradient, has an electric potential about 10 mV less inside than outside the plasma membrane, pumps cations in (K^+) and out (Na^+, Ca^{2+}), leaks cations in (Na^+, Ca^{2+}) and out (K^+), and has other cations and anions [H^+, Mg^{2+}, Cl^-, bicarbonate (HCO_3^-), biphosphate (HPO_4^{2-})] and H_2O crossing the membrane.

Table 3-3 gives some characteristics of the various mechanisms described above that are taken to be the causes of this dynamic control of water and solute concentrations inside and outside living cells.

3-2 TRANSIENT CHANGES IN TRANSPORT PROPERTIES

The dynamic but stable steady state characterized by a balance between pumps and leaks described in the last section applies quite generally to living cells. But many, if not all, cells exhibit reversible deviations from such a steady state. Such deviations can be described as changes in solute concentrations on the two sides of the membrane and in the electric potential

Table 3-3 Mechanisms of Transmembrane Transport

Mechanism	Driving force	Energy requirement	Coupled with	Membrane role	Significant parameter
Simple diffusion	Δc_s	No	—	Homogeneous barrier	P (for solute), L_p (for water)
Osmosis	Δc_s (or Δc_{H_2O})	No	—	Semipermeable barrier	L_p, σ
Simple current flow	$\Delta\psi$	No	—	Electric resistor (conductor)	P, z
Bulk flow (filtration)	ΔP	No	J_v (for solute)	Filter	L_p, σ
Facilitated diffusion	Δc_s	No	—	Carrier-containing barrier	Number of carriers, rates of reactions
Active transport	Metabolism (J_r)	Yes	ATP → ADP (?other)	Carrier, energy transducer	Number of pumps, rates of reactions
Flow-coupled transport	Δc_i	Indirect if any	J_i	Coupler, barrier	P_s, P_i, L_p (in some cases)

difference across the membrane. In human biology the outstanding examples are the plasma membranes of muscle and nerve cells. The extensively studied orderly changes in transmembrane potentials of particular muscle and nerve cells from particular invertebrate and vertebrate species are most often described in the terms and context of electric circuit theory (electrophysiology), because the observational methods are those of obtaining, amplifying, and recording electric potentials and currents from which are derived resistances and capacitances. As a result the transient membrane change associated with transmission of signals (for example, along a nerve axon[7]) is called an *action potential.* The steady state of constant potential (*resting potential*) is called the *resting state.*

The electric potential difference across the membrane of a nerve axon may be observed by recording from a pair of electrodes—one in the axon (in contact with the axoplasm) and one outside (in contact with extracellular liquid). The electric resistances of axoplasm and extracellular liquid are so small relative to that of the membrane that the observed currents and potentials are essentially those determined by the membrane. The resting potential observed in this way may be some 70 mV, with the extracellular fluid positive relative to axoplasm. If these electrodes are opposite one another across the membrane and close to the membrane, the potential difference during an action potential changes with time, as shown in Figure 3-6.

Study of the giant axons of certain invertebrates has shown that the transient change in potential is associated with transient changes in resistances of the membrane to transmembrane currents of Na^+ and K^+. The corresponding variations in Na^+ conductance[8] and K^+ conductance are shown in Figure 3-7 along with the time course of the action potential.

A variety of mechanical, chemical, and electric stimuli are capable of starting a propagated nerve impulse in an axon (as well as in a nerve cell body, nerve dendrite, and muscle cell). One way to initiate a propagated action potential is to sufficiently *depolarize* the membrane (that is, reduce

[7] The typical nerve cell consists of a *cell body* (cb) containing the nucleus, *axon* (a), and *dendrites* (d). The cytoplasm of the axon is called *axoplasm.*

[8] Conductance is the inverse of resistance. If L is conductance, Ohm's law may be written $i = \Delta\psi / R = L\ \Delta\psi$.

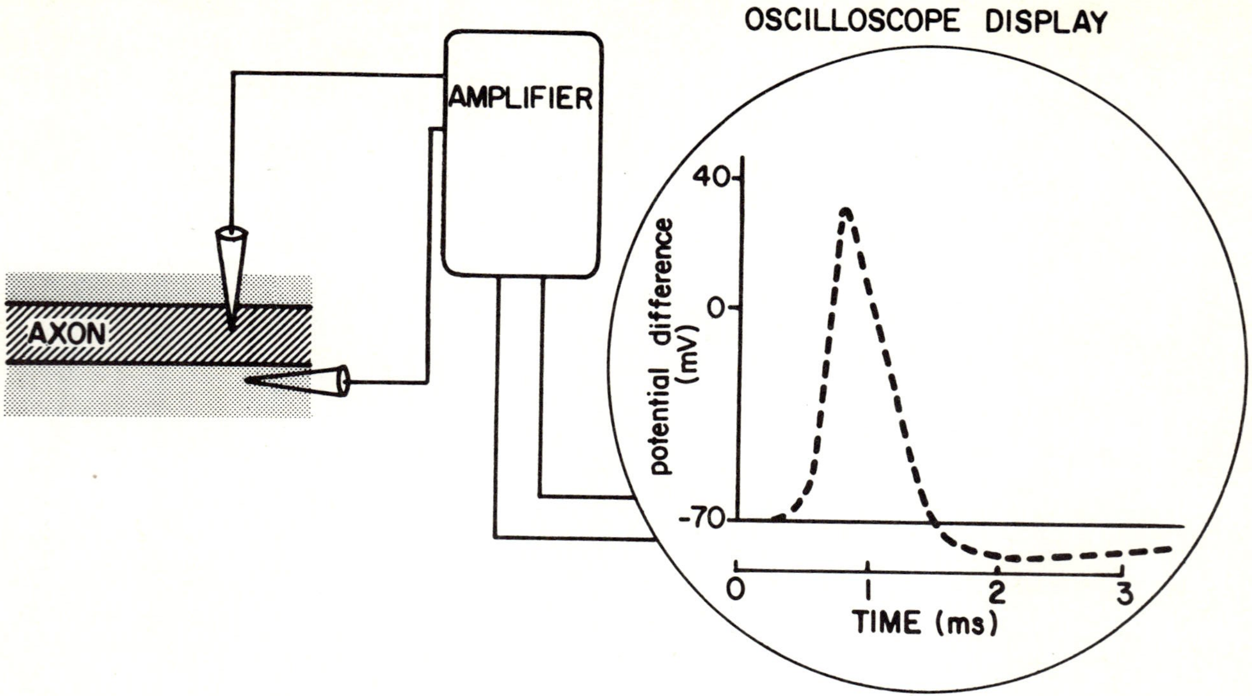

Figure 3-6 Experimental plan and record of an action potential of a nerve axon. The recording electrodes are placed as follows: one within the axoplasm (cytoplasm) and one in the extracellular medium. Initially the axoplasm is at −70 mV relative to the external medium.

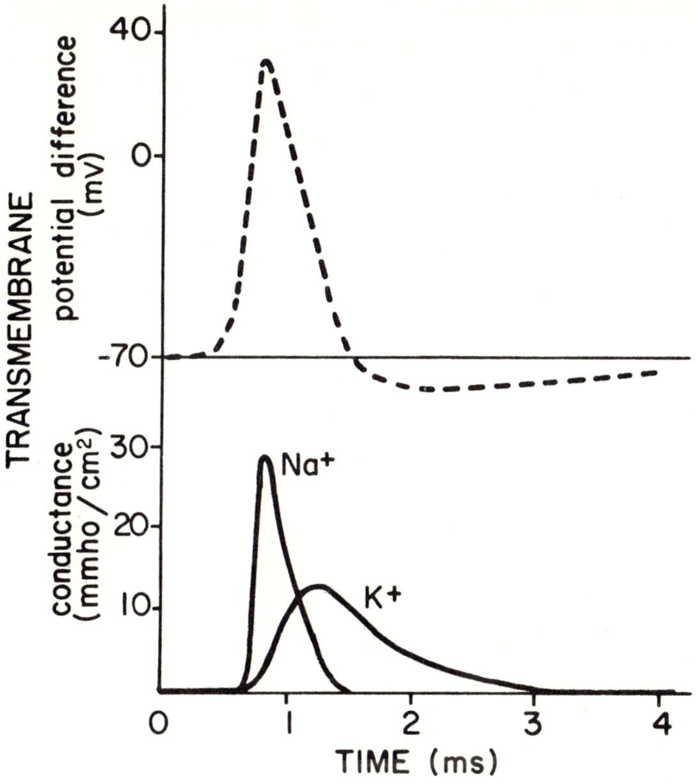

Figure 3-7 Temporal relations of Na^+ and K^+ conductances of nerve cell membrane to the action potential. The transmembrane potential difference (top) and the conductances (bottom) are shown on the same time scale.

the magnitude of the transmembrane potential difference). This triggers the usual conductance changes and thereby induces the action potential. The propagation of the nerve impulse may be explained as a progressive local depolarization of adjacent axon membrane. The volume conductors outside and inside the membrane have relatively low resistances; Na^+ and K^+ flow freely along parallel to the membrane. The K^+ concentration is lowered inside in the region of an action potential because the increased conductance is accompanied by an efflux down the concentration gradient. Similarly the concentration of Na^+ is lowered outside in the region of an action potential. The Na^+ effect is greater earlier in the course of the action potential. Therefore, the transmembrane potential difference is reduced (the membrane is depolarized) in advance of the action potential (toward the adjacent membrane in the resting state). Such a depolarization cannot occur behind the propagating impulse until the membrane is sufficiently repolarized. This period of insensitivity is called the *refractory period* and is long enough that propagation normally progresses in only one direction. The recovery to the resting state is accomplished by the continual operation of the ATP-energized one-to-one Na^+-K^+ pump during and after the conductance changes.

The process described above occurs in nerve and muscle cells (*excitable* cells), but the temporal length of the action potential and the length and relations of the changes in Na^+ and K^+ conductance vary widely among cells of an individual organism and among corresponding cells of different species.

Although the action potential itself is considered to be an all-or-none phenomenon of fixed form and magnitude, the sensitivity of a given cell membrane to stimulation can be altered both *in vivo* and *in vitro*. This alteration of sensitivity is an integral part of the function of more complex nervous systems. Action potentials in some cells can increase the sensitivity of a cell with which they interact—presumably by altering the state of its membrane—while other cells may decrease the sensitivity of the same cell. Thus the initiation of an impulse in a given neuron by a given stimulus will depend on the condition of the membrane as determined by the degree of *excitation* or *inhibition* by other neurons.

The fundamental questions here are: What molecular variations are associated with these variations in electric properties? How are such variations initiated? How are they controlled and reversed? Answers to these questions have been pursued both experimentally and theoretically. Experimentally, some probes of molecular conformation and motion have revealed variations in concert with changes in transmembrane potential difference during action potentials. Also, specific membrane components, proteins and lipoproteins, have been found to bind substances, such as acetylcholine, which alter the membrane conductivity and electric potential. On the theoretical side such observations have been used as the basis for hypotheses that plausibly relate the action potential to molecular events in the membrane. There is good reason to expect a satisfactory explanation of the known behavior of excitable membranes on the basis of known principles of molecular physics and protein chemistry.

The connection with molecular events is of great interest in the study of the effects of hormones and drugs that involve transient conductance changes. A hormone or drug may change a biomembrane in a variety of ways, but often the physiologically significant change is in its transport properties. For example, the hormone vasopressin (antidiuretic hormone) in several mammals changes the permeability of the collecting duct epithelium of the kidney to water to a value several times larger than when the hormone is absent. Presumably, this is a specific effect on the plasma membrane of the tubular epithelial cells. Estrogen increases the transport of amino acids into the musle cells of the human uterus. Progesterone changes ionic gradients across the plasma membrane of uterine muscle cells and inhibits their response to another hormone, oxytocin. Many other membrane effects of hormones have been observed.

With respect to drug action the membrane is often the significant target. Curare and its artificial analogs specifically block the receptor sites of the muscle cell that initiate impulses in response to a normal chemical stimulus from the nerve ending. Topical anesthetics reduce the excitability of the plasma membrane of pain nerve fibers.

The relation between chemical modification of membrane and action potential is clear in synaptic transmission. The synapse is a specialized relationship between two neurons or between neuron and muscle cell. One of these cells, the presynaptic neuron, is the sender, and the other, the postsynaptic neuron or the muscle cell in the case of the neuromuscular junction, is the receiver. The process of synaptic transmission illustrates the chemical stimulus of an excitable cell mentioned above and the membrane-synthesizing secretory process, which is discussed in Chapter 5.

The arrival of a nerve impulse at a cholinergic synapse is followed by the release of acetylcholine from the terminal processes of the presynaptic axon into the synaptic cleft, some 20 nm across, between the presynaptic and postsynaptic neuronal cell membranes—or between presynaptic axon and motor end plate of the muscle cell in the case of the neuromuscular junction. This acetylcholine is synthesized in the presynaptic axon and confined in secretory vesicles formed of closed biomembranes. The release into the synaptic cleft, which is triggered by the arrival of an action potential, is accomplished by fusion of the secretory vesicle membrane with the cell membrane, thus opening the vesicle into the synaptic cleft (Figure 3-8). The acetylcholine diffuses across the gap and binds to specific protein or lipoprotein receptors in the plasma membrane of the postsynaptic neuron. The interaction of acetylcholine with receptor initiates an action potential that is propagated over the postsynaptic neuron or muscle cell. This process is a strong indication that a thorough understanding of biomembranes will reveal common mechanisms for such seemingly different phenomena as electric stimulation of excitable cells and chemical (drug, hormone, and neurotransmitter) effects on cell behavior.

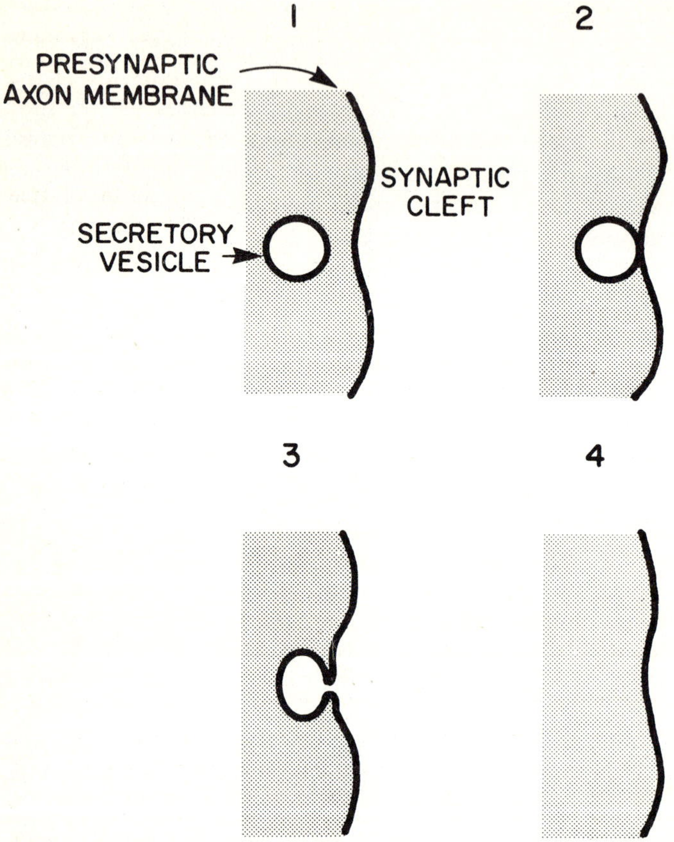

Figure 3-8 Neurotransmitter release through presynaptic cell membrane. The secretory vesicle contacts the plasma membrane (2), fuses (3), and releases its contents into the synaptic cleft.

3-3 SORPTION THEORY

The preceding sections have presented membrane theory, which holds that the asymmetric distributions of water and solutes about biomembranes is explained by properties of the membrane alone and known laws of the physics and chemistry of solids and solutions. This explanation has been challenged on the grounds that (1) the data that it explains are not a true representation of the situation; and (2) if the data were correct, the explanation would require a greater expenditure of energy to maintain the observed concentration differences than that expended by the cell in question.

Consider the problem of determining the extracellular and intracellular concentrations of Na^+, for example, in a suspension of lymphocytes. One centrifuges the suspension, removes an aliquot of supernatant, and determines the Na^+ concentration by flame photometry. To determine the intracellular concentration one washes the cells by resuspending them and centrifuging once more to remove Na^+ from the surface of the cell membrane and to reduce the Na^+ trapped between cells. However, washing the cells in Na^+-free medium causes a loss of Na^+ from within the cells. To get intracellular Na^+ into solution for flame photometry, one then resuspends the cells in fresh Na^+-free medium and breaks them up with ultrasound, centrifuges the fragments, precipitates the protein from the remaining solution and makes the measurement. But that measurement is made after losing Na^+ in the wash and freeing Na^+ bound to protein and from intracellular organelles. To study the transmembrane transport of Na^+ this value is used as the intracellular cytoplasmic concentration; the difference between Na^+ dissolved in water—not including Na^+ bound to proteins and organelles—is the significant gradient. Needless to say, many doubts may be raised. More direct measurement of activities can be made, and isotopic methods eliminate some problems, but some questions always remain.

The second point—energy balance—is still hotly contested. There has been a thorough and thought-provoking effort to measure the steady-state utilization of metabolic energy by the bacterium *Escherichia coli* in a basal stationary phase where the cells maintain transmembrane gradients without growth or net synthesis of cell components. This energy expenditure was found by one group to be 15% of the energy computed on physicochemical grounds to be required to maintain the dozen or so gradients thought to represent all the major transmembrane concentration differences as determined in the usual way in the context of membrane theory. The experiments and arguments used in arriving at this conclusion have been seriously questioned, and experiments suggesting the opposite conclusion have been done with *E. coli* and with erythrocytes. The issue is, to say the least, unresolved.

The alternative view of distribution of solutes in cells, referred to here as the *sorption theory* (related to the *association-induction hypothesis*), is based on the well-known physiochemical phenomenon of *adsorption*. Adsorption is a process of noncovalent interaction between a solute and a macromolecule or complex that binds it (for example, between an inorganic ion and a protein) so that their combined free energy is lower than it is in the unbound state (for example, the ion dissolved in the aqueous medium). Such an interaction may occur between a small solute and a macromolecule—a colloidal particle—or between a small solute or macromolecule and a biomembrane or other surface. For example, Na^+ can associate with a protein, in solution or fixed to or in a biomembrane, in an electrostatic interaction so that an equilibrium exists between Na^+ in solution and bound Na^+. The relative amounts of Na^+ bound and in solution will depend on the difference in free energy of the two states.

Such adsorption may occur with water molecules. Vicinal water is a result of this process. Nonelectrolytes and polyelectrolytes are also adsorbed. Adsorption of a macromolecular polyelectrolyte such as a protein may produce a gel along surfaces, which in effect creates a distinct aqueous compartment in its interstices and also binds small solutes. Such situations are invisible to a method of concentration measurement such as that described above for intracellular Na^+. The actual concentration in the cytoplasm of a cell may be very different from the estimate based on the assumption of a homogeneous intracellular phase occupying the total volume of water in the cell. The estimate may be either too high or too low, because either solute (Na^+, for example) or solvent (water, for example) may be adsorbed.

Some of the more recently developed techniques for measuring states of molecules with respect to interactions with other molecules have produced evidence consistent with considerable binding of Na^+, K^+, H_2O, and other components of cytoplasm. There is little question that adsorption occurs in living cells; the issues are the extent to which it occurs, and the relative merits of the sorption theory and membrane theory as explanatory and predictive models. The committed membrane theorist claims that only better data and a refined membrane theory are needed. The sorption theorist says that those huge gradients do not really exist at all, that one needs to reinterpret the data, add some observations based on the sorption theory, and thereby relieve the membrane of the impossible demands of the membrane theory.

The problem of transient and reversible changes such as those discussed in Section 3-2 is perhaps a more severe test for both viewpoints than is the steady state. Here the membrane theorist is still unable to explain the observed phenomena with changes confined to the biomembrane, although some data obtained with nondisruptive physical methods during action potentials and during interactions with drugs support this theory. The sorption theorist, also backed by observations, invokes changes in the intracellular proteins, which produce rapid changes in binding properties and which shift the preference of binding sites from one entity to another—Na^+ to K^+, for example.

One may hope that the argument between sorption theory advocates and membrane theory advocates, together with the thoughtful evaluation of those simply interested in understanding biologic systems, will teach us much.

3-4 SUMMARY

The details of transmembrane transport determine the widely varied contents of cellular and extracellular compartments. The mechanisms of transport may be classified as follows:

1 Diffusion; flow proportional to concentration gradient.
2 Bulk flow; proportional to pressure gradient.
3 Osmotic flow; proportional to osmotic pressure that is, in turn, proportional to concentration gradient of solute for a given reflection coefficient.
4 Component-reactive transport; characterized by substrate interaction with specific membrane component and including (a) facilitated diffusion (flow proportional to concentration gradient), and (b) active transport driven by energy-yielding metabolic reaction.
5 Flow-coupled transport; proportional to flow of solvent or another substrate.

Such processes maintain steady-state differences between solutions on the two sides of a biomembrane and, in specialized cells and in cells responding to hormones and drugs, undergo ordered, transient, and reversible changes. Adsorption of water and solutes onto intracellular proteins may account for some of the apparent differences between solutions separated by membranes.

Chapter 4

Cell Compartmentation and Membrane Asymmetry

Perhaps the simplest, most obvious achievement of the plasma membrane is to keep the larger components of cells (other organelles and macromolecules) together. But the discussion of transport in the last chapter and the phenomena described in the following chapters are clear indications that biomembranes are much more than mere containers of large things. Nevertheless, the idea of simple containment is an important part of our understanding of cell biology. Also, it applies to biomembranes other than the plasma membrane.

The asymmetry of biomembranes–that is, the differences between opposite surfaces–is evident in the distribution of phospholipids and cholesterol, the orientation of intrinsic proteins, the location of extrinsic proteins, and the one-sided location of carbohydrates constituting the glycocalyx. These asymmetries were discussed in Chapter 2. The discussion of transmembrane transport (Chapter 3) is actually a discussion of some functional consequences of membrane asymmetry; the differences in constitution of the two media adjacent to a biomembrane seem clearly related to the specific orientation of proteins in the membrane. For example, receptors for Na^+ and K^+ are on opposite sides; the Na^+-K^+ ATPase of the plasma membrane releases phosphate specifically at the cytoplasmic interface; the movement of Na^+ and K^+ across the membrane in opposite directions is associated with enzyme-catalyzed phosphorylations. The Na^+-K^+ transport system clearly demonstrates the interplay between the membrane's barrier and directional properties.

It is impossible to explain the membrane-related phenomena of biologic systems on the basis of barrier properties alone or transmembrane asymmetry alone. Nevertheless, in trying to understand the general properties of biomembranes it is convenient to separate these properties with respect to their general effects. Section 4-1 is a reminder of the significant consequences of biomembranes as barriers. Section 4-2 is a discussion of membrane-bound enzymes as a major example of the asymmetry of biomembranes. In Section 4-3 the intricate interplay of these effects is demonstrated in a discussion of oxidative phosphorylation in mitochondria. Section 4-4 presents some generalized dynamic properties of a system of biomembranes that result in asymmetry across multimembrane structures.

4-1 EFFECTS OF SEGREGATION

The activities of a living cell are strongly coupled and thoroughly integrated. Specific variations in different types of cell require juxtaposition of components in a highly organized way. The plasma membrane keeps the nucleus close to the endoplasmic reticulum, which synthesizes specific proteins on instruction from the nucleus. It also keeps the mitochondria, which supply most of the energy for cell functions, close to the energy-requiring processes. This holds true for all organelles except perhaps secretory and storage vesicles, peroxisomes, and lysosomes, which do however require other organelles and their products in order to be activated to perform their functions.

This simplest form of segregation–joint confinement of functionally related organelles–is the defining feature of the eucaryotic cell.[1] The genetic apparatus is segregated from other cell components by the nuclear envelope, a double biomembrane system with complex structures, the nuclear pores, which tack the inner and outer nuclear membranes together and apparently govern the passage of certain macromolecules between nucleoplasm and cytoplasm. This nuclear envelope, it seems, separates the information storage and basic control apparatus of the cell from the energy-producing and major synthetic and packaging apparatus. Through a system of repressors and

[1] The association among organelles is more specific than simple confinement within the plasma membrane, as shown in Chapter 5.

derepressors particular sequences of genes can, in response to the environment, be activated to produce messenger ribonucleic acid (mRNA). This mRNA passes through the nuclear envelope, presumably via the nuclear pores, to induce the synthesis of specific proteins in the endoplasmic reticulum. This process depends on the juxtaposition of macromolecular components of the system. The segregating membrane ensures this juxtaposition and prevents mixing with the extranuclear synthetic apparatus, which produces many copies of many different proteins with the same apparatus according to varied signals (mRNA) from the nucleus. Similarly the process of replication of deoxyribonucleic acid (DNA) prior to cell division requires the interaction of existing DNA with enzymes catalyzing the synthesis and polymerization of components of the new DNA.

Confining material inside a closed sac protects both contents and surroundings. In the cases of lysosomes and peroxisomes, the biomembranes package enzymes that could catalyze the breakdown of essential cell components. The hydrolytic enzymes of lysosomes and the oxidative enzymes of peroxisomes are damaging, probably lethal, to the cells that contain them. With the vesicle membrane intact the enzymes are conserved and the potential substrates outside the vesicle are protected. It appears that these contained enzymes are brought into play when the containing organelles fuse with phagocytic or pinocytotic vesicles (compare Section 6-4). The membrane fusion process, which seems to bring the contents into action, is not understood, but it may be expected that the asymmetry of the vesicle membranes and receptors on the cytoplasmic interface will be found to mediate the transition from storage to activation (release) of vesicle contents. These membranes, then, are both reservoir and activator for the vesicle contents.

Other organelles may well constitute cases of protection of contents from cytoplasmic enzymes and other constituents. Such a protective role might be one of simply avoiding interference in a reaction catalyzed by a relatively nonspecific enzyme (for example, in ester formation). A good example is the secretory vesicle that stores cell products behind a protective barrier.

The activity of a reactant is a reflection of the probability of a sequence of events. The probability that a reaction occurs in some short time interval is the product of (1) the length of the interval, (2) the probability per unit time of reactants making contact in that interval, (3) the probability of a required mutual orientation, and (4) the probability of reaction given the required contact and orientation. The probability of reactants making contact at a given temperature in a given time interval is proportional to the product of the concentrations of the reactants.[2] There are two ways

[2] The concentration (or activity) of a reactant is not constant if the reaction is

to increase concentration: increase the number of molecules; or decrease the volume of distribution. Segregation of an enzyme and its single substrate into an organelle of 20 μm^3 in a cell of 400 μm^3 increases the probability of contact in an interval of time 20 × 20 or 400 times. If this were the rate-limiting step in the production of a secreted substance, then the cell with such an organelle could secrete at the same rate as a cell without this compartmentation with only 0.25% as much enzyme in the presence of the same substrate concentration. An enzyme-catalyzed bimolecular reaction would be 20 × 20 × 20 or 8000 times more probable in a given small time interval. This is an oversimplication, but it correctly suggests the significance of joint confinement of reactants that participate in a single reaction sequence or in related reactions.

In addition to increasing the concentrations of reactants, compartmentation by membranes may modify the activity of a reactant by other changes in the medium that affect the probabilities involved in the reaction. Such changes may be in pH, ionic strength, or other factors that affect the shape or hydration (and thereby the translational and rotational diffusion coefficients and reaction probability) of the reactant molecule and/or enzyme.

Another effect of segregation is pulse release of the segregated material. In Section 3-3 the mechanism of synaptic transmission was mentioned. In that case secretory vesicles contain a neurotransmitter. These chemicals are stored, inactive, until the membrane changes associated with arrival of a nerve impulse trigger the fusion of vesicle membrane with presynaptic nerve cell membrane. A pulse of transmitter then is released into the synaptic cleft where it interacts with receptors on the postsynaptic membrane (or muscle end plate). This achieves a sudden large increase in concentration that is transient. (In the neurotransmitter case, it is transient because diffusion and enzymatic inactivation quickly dispense with neurotransmitter in the cleft.)

Another example of pulse release occurs with secretory cells such as the pancreatic acinar cell that releases, in response to hormonal signals, enzymes into the pancreatic duct system. This makes available a larger amount of enzyme in a shorter time than expected if the enzymes, rather than being stored, had to be synthesized *de novo*.

The effects of segregation, then, may be categorized as follows: juxtaposition of functionally related cell components; protection and storage of cell products; and control of activities of reactants and catalysts. The

---

occurring. The requirement that the time interval be short removes this objection by ensuring constant probability densities (probabilities per unit time) in the interval. To be precise, one must use the limit as the interval approaches zero.

processes of the living cell are an interplay of these effects, but the categories are conceptually useful.

4-2 MEMBRANE-BOUND ENZYMES

The coincident development of electron microscopy and ultracentrifugation in the middle decades of this century—along with biochemical methods for identifying constituents of biologic materials—gave biologists the means to separate and characterize a variety of distinguishable organelles and biomembranes. One of the biochemical methods applicable to isolated membranes is that of measuring enzyme activity. Almost all biomembranes have been found to have enzymatic activity associated with them, wherever it has been looked for thoroughly. The relatively simple human erythocyte has over 50 distinct enzymes associated with its only membrane.

The first question to be answered when interpreting such findings is whether the activity is truly intrinsic to the membrane or merely a result of contamination of the preparation with enzyme from another cell fraction. This question takes us back to the issue raised in Section 1-3, the matter of definition. In fact, differences occur among scientific papers as to whether a particular catalytic activity results from a membrane-bound enzyme or from an enzyme that contaminates the membrane preparation. It is not enough to say that the enzyme is, by operational definition, membrane-bound in the case of the procedure that shows it. As will be seen in Section 4-3, the issue is more significant to understanding organelle function than such a facile answer suggests. For our purposes here, it can be said with conviction that there are many genuinely membrane-bound enzymes in many different membranes. Their activities are associated with proteins that fit the criteria of intrinsic membrane proteins, and the reactions they catalyze fit into established sequences associated with membranes and with transmembrane differences in reactant and product concentrations. In addition, many of these membrane-bound enzymes are influenced with respect to enzymatic activity by their association with membranes or with constituents of membranes.

In view of what has just been said, the term *membrane-bound enzyme* may be misleading. Such enzymes are better regarded as integral parts of their respective membranes rather than simply bound to membranes. It cannot be overemphasized that many membrane-bound enzymes are thoroughly integrated into the structure of the membrane and that the behavior of such an enzyme—the behavior of the membrane—is inseparable from the relations of the enzyme and its parts to the rest of the membrane and its parts. Figures 4-1–4-5 illustrate the topological characteristics involved and may clarify the following discussion.

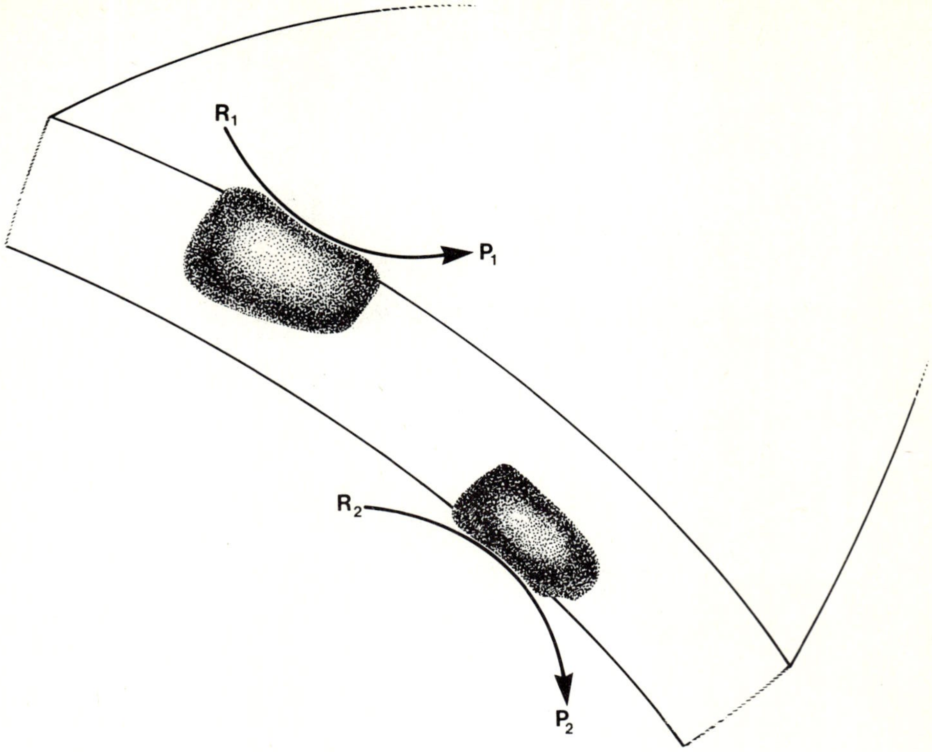

Figure 4-1 Membrane-bound enzymes catalyzing reactions on one side of the membrane. R_1 and R_2, reactants; P_1 and P_2, products.

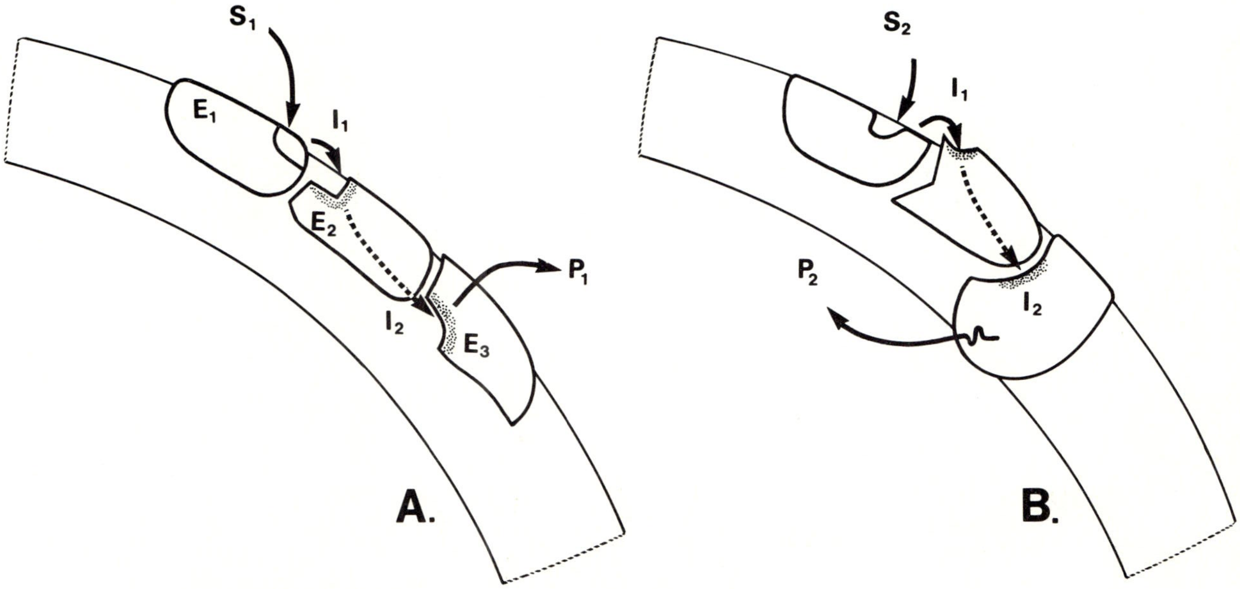

Figure 4-2 Hypothetical sets of membrane-bound enzymes catalyzing sequential reactions. E_1, E_2, and E_3, enzymes; S_1 and S_2, substrates; P_1 and P_2, products; I_1 and I_2, intermediate product-substrates. (A) Product appears in same compartment as substrate. (B) Product appears on opposite side of membrane.

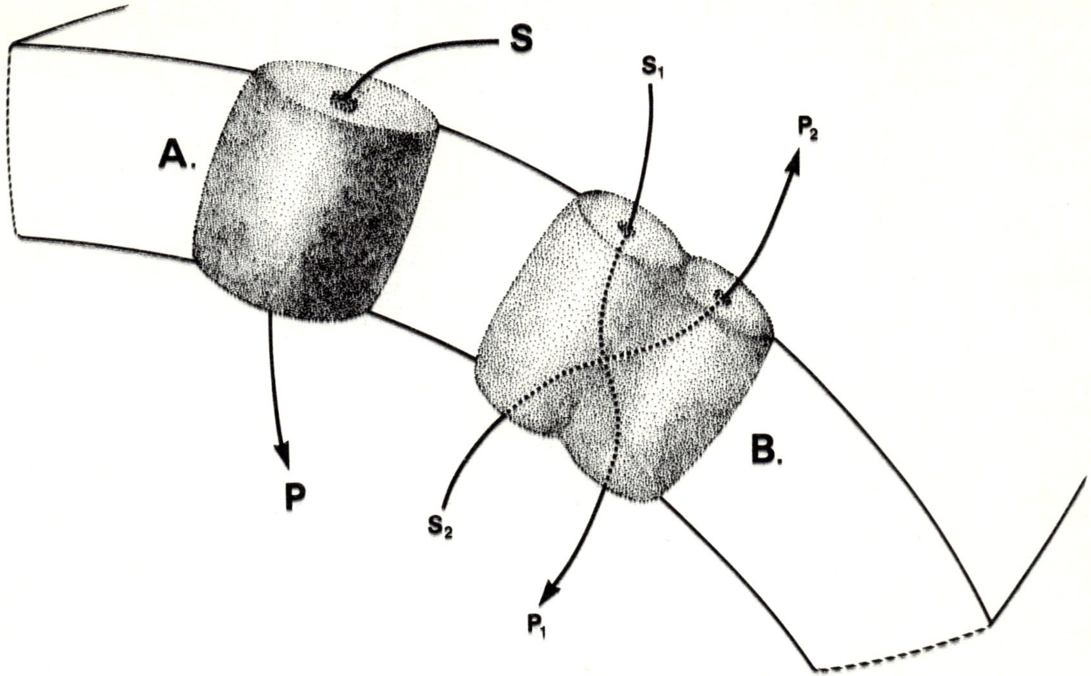

Figure 4-3 Membrane-bound enzymes (hypothetical) that separate substrates and products. (A) The reaction is associated with transmembrane displacement; if P is chemically identical with S, this is component-reactive transport. (B) A complex that couples reactions and displacements; if P_1 and P_2 are chemically identical with S_1 and S_2, respectively, this is exchange transport.

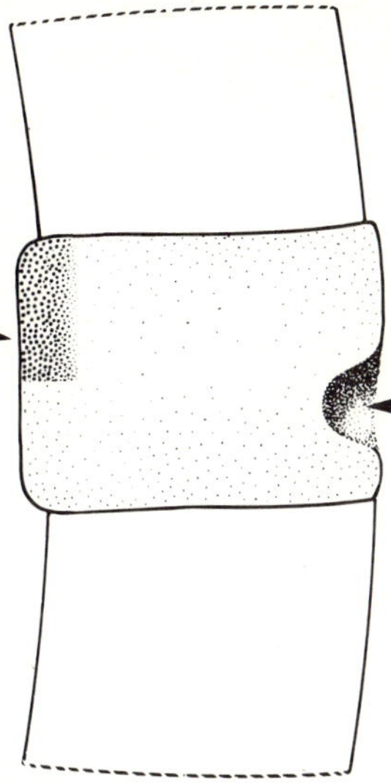

Figure 4-4 Membrane-bound enzyme with allosteric and active sites exposed to different aqueous media (hypothetical).

There are several ways in which the fixation of an enzyme in a biomembrane can significantly alter the behavior of the enzyme. These have to do with reactants and products, on the one hand, and with the state of the enzyme itself, on the other hand. Furthermore, the nature of the enzyme may be coupled to the distribution of reactants by the membrane. For example, an enzyme may be incorporated into a membrane in such a way that its binding site for a particular substrate is accessible to the aqueous medium on one side of the membrane only. Again, the Na^+-K^+ ATPase system is an example. This specificity of direction is sometimes referred to as the *vectorial* nature of a biomembrane or of a membrane-bound enzyme. Its importance stands out in the discussion of the mitochondrion in Section 4-3.

The effect of compartmentation of substrate concentration, mentioned above, applies to reactions catalyzed by membrane-bound enzymes as well as to reactions occurring entirely in solution. However, in the case of an enzyme that is part of a membrane the orientation factor in reaction probability is radically changed. In addition, there is the possibility that the reactants and products can be separated by the membrane or that they may be fixed at different sites on the same side of a membrane. The latter seems to occur where there is a group of membrane-bound enzymes associated with a sequence of reactions–such as those of respiration described in Section 4-3–in which the product of one reaction is a reactant in the

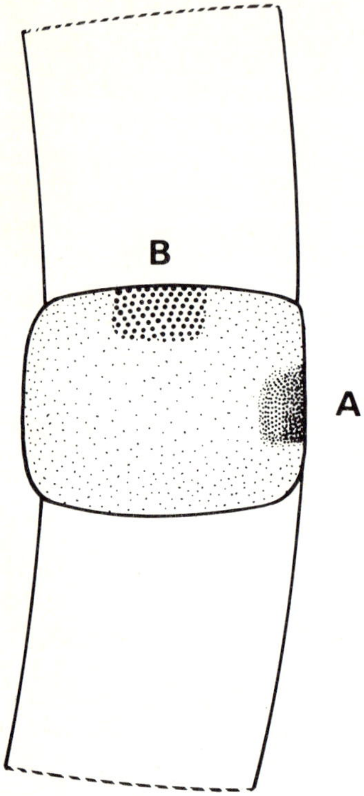

Figure 4-5 Hypothetical situation in which the active site A of an enzyme or enzyme complex is exposed to an aqueous medium and the allosteric site B is exposed to a hydrophobic environment.

next. In this case, the binding site of a membrane-bound enzyme serves to fix the product to the membrane; this product then becomes reactant in another enzymatic reaction (Figure 4-2). (There is an interesting parallel in the sequences of loci along DNA molecules, that code for enzymes of reaction sequences, where the respective sequences correspond in order.) When the product-reactant concentration is locally increased by holding the substrate within a certain volume, the effect is to speed up a reaction sequence carried out by a limited number of substrate and catalyst molecules.

Another kind of segregation of reactants and products occurs where the membrane-bound enzyme spans the membrane, or where the complex of which it is a part spans the membrane (Figures 4-2–4-5). In some such cases the reactants are bound by the enzyme at one interface of membrane with aqueous medium, and the products are released at the opposite interface. The case of facilitated diffusion discussed in Section 3-1 may be thought of as an enzyme reaction where the reactant and product are identical except for being in different compartments (and possibly at a different electrochemical potential). In the case of bacterial membranes, such enzymes (complexes) have been called *permeases*. The Na^{+}-K^{+} ATPase system, apparently a complex of separable enzymes with different receptor sites on different sides of the membrane, is a more complex example of such a system and one that involves greater free energy differences. It constitutes an exchange transport system, because each ion is a required reactant in the transport of the other.

Transmembrane segregation of reactants and products of an enzyme-catalyzed chemical reaction occurs in some cases so as to combine a metabolic step and a transport step. For example, sucrase is an enzyme of the membrane of intestinal mucosal cells. It hydrolyzes the disaccharide sucrose into the monosaccharides fructose and glucose. The glucose product of the reaction does not appear on the luminal side of the membrane where the sucrose had been; rather it is found in the cytoplasm of the mucosal cell. Either the sucrase spans the plasma membrane and serves a combined hydrolase and permease function, or it is closely associated with some transport device that binds the glucose, before it is lost to the lumen, and transports it across the membrane.

Synthesis and segregation seem to occur simultaneously in the synthesis of protein by some ribosomes. Ribosomes mediate the RNA-guided polymerization of amino acids to form proteins with specific primary structures. Some ribosomes are found bound to membrane. These ribosomes are discussed in Chapter 5 in connection with the biogenesis of membranes. Here their role in the synthesis of proteins that are contained by organelles and secreted by cells is pertinent. A ribosome bound to the outside—the cytoplasmic side—of a closed intracytoplasmic membrane has access to amino acids (aa) and transfer RNA (tRNA) and to the aa-tRNA complex in the cytoplasm. In some cases, such as albumin produced in liver parenchymal cells, the synthesized protein appears exclusively on the inside—the side opposite the cytoplasm—of the membrane. At least the evidence strongly suggests this. If true, this is an example of segregation combined with a many-step synthesis—the formation of peptide bonds sequentially according to the genetic instructions—mediated by a sort of membrane-bound enzyme.

There are several possible effects of an enzyme's being a part of the membrane that are related to the state of the enzyme itself. As intrinsic proteins, membrane-bound enzymes have more than just an aqueous environment. They interact with other constituents of the membrane, in some cases with other proteins, but in all or most cases with lipids. In general membrane-bound enzymes probably interact with both polar and hydrophobic surroundings. Indeed, the enzymatic activities of many depend on contact with lipids. This contact with both aqueous and lipid regions makes possible access to both water-soluble and lipid-soluble substrates. It also becomes possible that the active site and the allosteric site of a single enzyme can be in contact specifically with substrates and/or cofactors that differ in their affinities for water and lipid (Figure 4-5).

In the case of enzymes spanning the membrane, different parts of the enzyme can be in contact with aqueous media at different pHs, different ionic strengths, and different viscosities. These contacts with different media and with membrane constituents influence the conformation of the enzyme.

There is considerable evidence that activities of membrane-bound enzymes are dependent on their being part of the membrane. Some enzymatic activities found in isolated membranes survive the procedures used to separate intrinsic proteins, but often a specific enzymatic activity is reduced or completely lost when the protein associated with it is separated from the lipid components of its membrane. With some such isolated enzymes it is possible to enhance the activity by adding the appropriate lipid or lipids to the isolated enzyme—sometimes to the point of fully restoring the activity to what it was in the membrane preparation. The dependence on lipid varies among membrane-bound enzymes with respect to specificity. Cholesterol may suffice for one; for another, cholesterol may do less well than phospholipid. Furthermore, the efficacy of phospholipids depends in some cases on the polar head group: choline phosphoglyceride, but not ethanolamine phosphoglyceride, may work. In some cases the nature of the fatty acid groups is significant. From the limited number of studies of this effect, it may be suggested that unsaturation and shortness of hydrocarbon chains make a phospholipid more effective in restoring enzymatic activity (these are properties that tend to increase the fluidity of membranes, see Section 2-2). In view of these influences one may assign to the membrane the role of a cofactor in the regulation of enzymatic activity.

Having the foundation laid above, one can appreciate the beautiful example of interaction of barrier properties with coupled vectorial membrane enzymes in contact with aqueous media and lipid regions given in the following section.

4-3 THE BIOMEMBRANE AS AN ENERGY COUPLER

The vectorial nature of membrane-bound enzymes and the fact that enzyme-catalyzed reactions are the major means of transferring energy to usable and storable forms suggest fascinating possibilities for coupling transmembrane gradients to chemical reactions. Indeed, as mentioned in Section 3-2, the Na^+-K^+ exchange mechanism that normally transports both Na^+ and K^+ across the membrane from a lower to a higher electrochemical potential at a cost of ATP can, in the case of the erythrocyte, be run backward by abnormally large gradients. This produces the higher-energy ATP from the lower-energy ADP and phosphate. An analogous process is described below as an explanation of ATP production by mitochondria. This description of mitochondrial function is known as the *chemiosmotic theory*.

The organelles that are the primary energy converters of eucaryotes are chloroplasts in the case of plants and mitochondria in the case of animals. These organelles produce ATP; their membranes are central to the process. In procaryotes, such as the bacterium *E. coli*, the cell membrane itself is associated with ATP production. It may well be that the early evolution of eucaryotic cells incorporated a formerly free-living form that became the mitochondrion in the animal line and the chloroplast in the plant line.

The mitochondrion of animal cells is an organelle surrounded by cytoplasm. Structurally it is composed of a closed, smoothly curving outer membrane and a closed, more or less convoluted, inner membrane. The convolutions of the inner membrane form the *cristae* (Figure 4-6). There are associated with the mitochondrion three aqueous compartments: the cytoplasm, the *intracristal space* between the two membranes, and the so-called *matrix* inside the inner membrane. Both these membranes are biomembranes as defined in Section 1-3; they are continuous, closed, and composed of a lipid bilayer and embedded protein. For an illustration of a biomembrane as a vectorial catalytic partition between compartments, attention should be focused on the inner membrane and the two distinct aqueous regions adjacent to it—the intracristal space and the matrix.

The inner mitochondrial membrane encloses within the matrix the substrates and enzymes associated with the tricarboxylic acid (TCA) cycle that produces the reduced form of nicotinamide adenine dinucleotide (NADH) and H^+ from NAD and the energy derived from metabolism of carbohydrates. The ultimate form of this energy, used in a variety of processes by the cell, is ATP. The transformation from NADH + H^+ to ATP depends on the inner membrane as both a barrier and a directional rack for specific membrane-bound proteins. The events mediating this transformation are represented in Figures 4-7 and 4-8.

One of the proteins bound to the inner membrane is a flavoprotein (FP), the active component of which is flavin mononucleotide (FMN). FMN undergoes the following reaction to form the reduced form $FMNH_2$:

$$\mathrm{FMN + NADH + H^+ \rightarrow FMNH_2 + NAD^+}$$

The flavin group is a heterocyclic compound that undergoes the oxidation-reduction reaction:

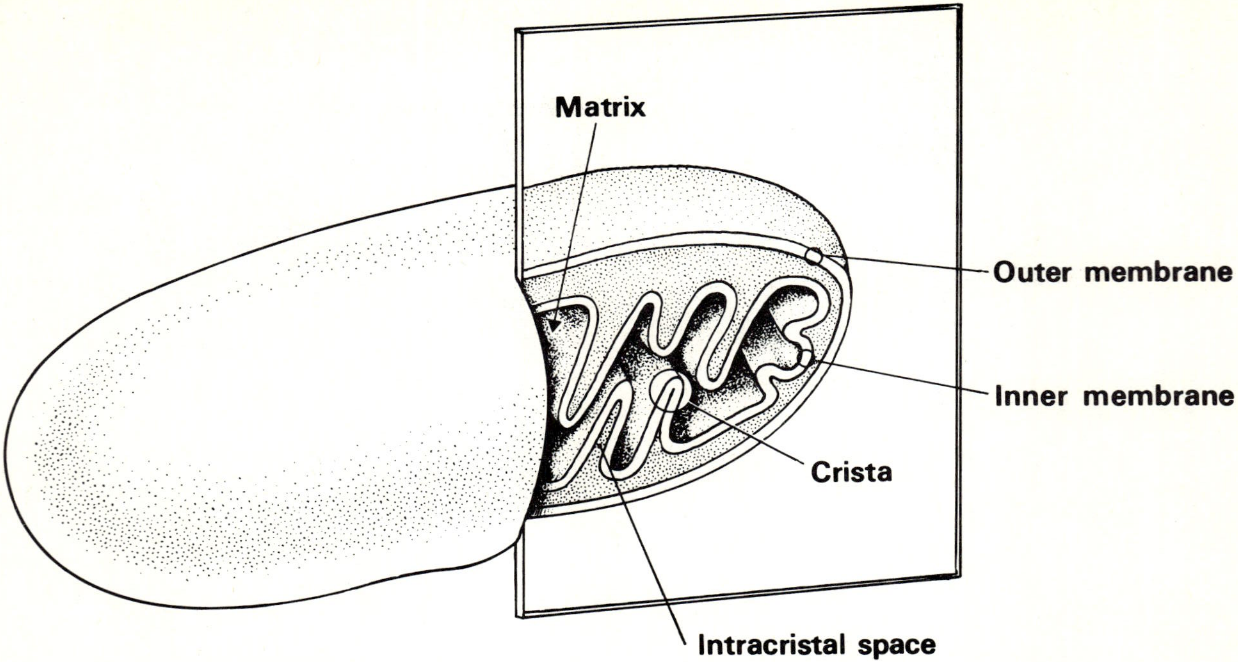

Figure 4-6 A mitochondrion. The two membranes form closed sacs that divide the region into three compartments: matrix, intracristal space, and cytoplasm.

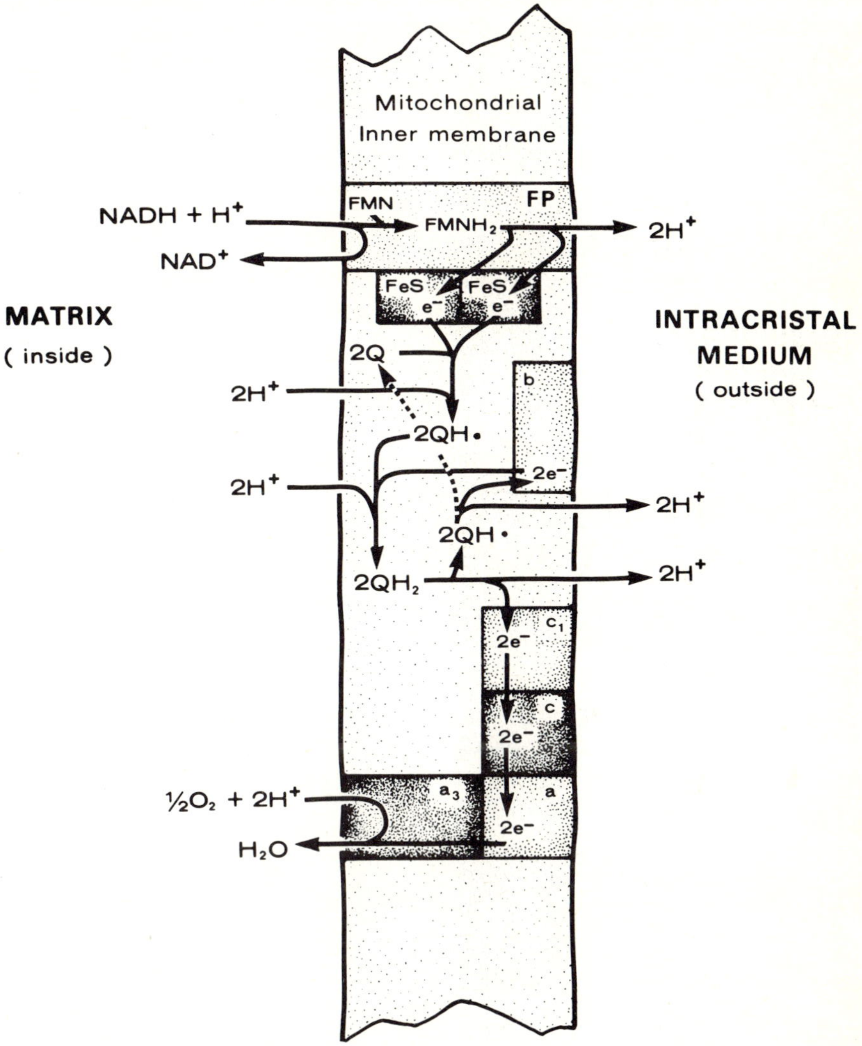

Figure 4-7 Representation of the locations and actions of the proton-gradient-generating components of the inner mitochondrial membrane. The symbols are defined in the text.

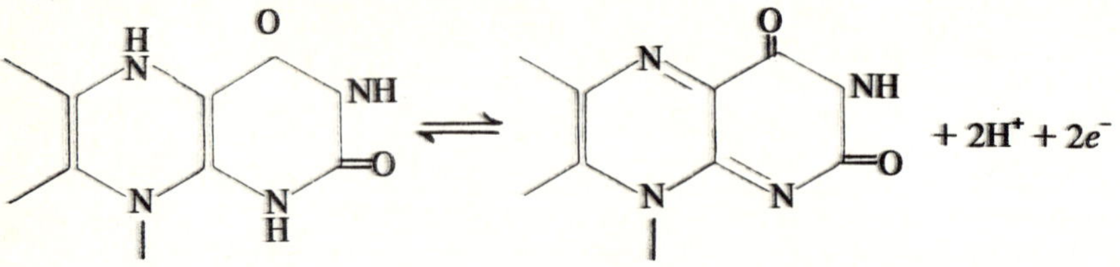

In the reaction above, NADH and H^+ donate two protons and two electrons. FP is situated so that one NAD^+ appears in the matrix solution and two protons appear in the intracristal solution *outside* the membrane, following the reaction:

$$FMNH_2 \rightarrow FMN + 2H^+ + 2e^-$$

The two electrons from FP are accepted by membrane-bound proteins called iron-sulfur proteins (FeSP). Evidence for the sequence of events

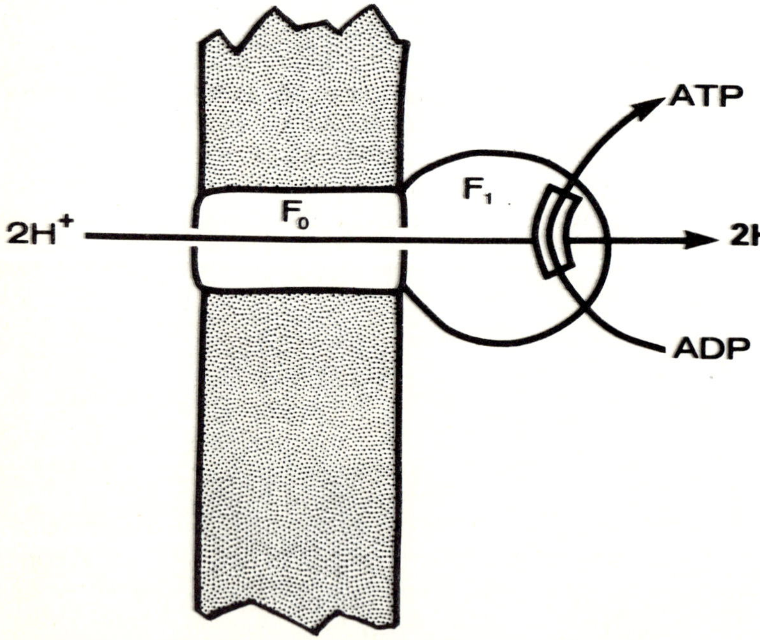

Figure 4-8 Representation of the protein complex coupling the translation of protons with the generation of ATP. F_0 is embedded in the inner mitochondrial membrane; F_1 is exposed to the matrix. Translation of two H^+ is associated with formation of one ATP in the matrix.

proposed to carry these electrons to the point where they reduce H^+ and O_2 to produce water is indirect and subject to question. The scheme presented here, however, is consistent with the evidence and reflects the role of the membrane, which is well established in general respects. In particular, the chemical intermediates are consistently found, and the evidence for the gradients across the membrane and for the necessity of the gradients for ATP production is well established.

Electrons move from FeSP to participate in the following reaction:

$$Q + H^+ + e^- \rightarrow QH\cdot$$

The dot represents an unshared valence electron of an oxygen atom, and the symbol Q represents the organic molecule ubiquinone, which is lipid-soluble and possibly quite mobile in the membrane. Like the flavin group and its N atoms, the quinone ring of Q can accept and give up electrons and protons at the two O atoms in two steps:

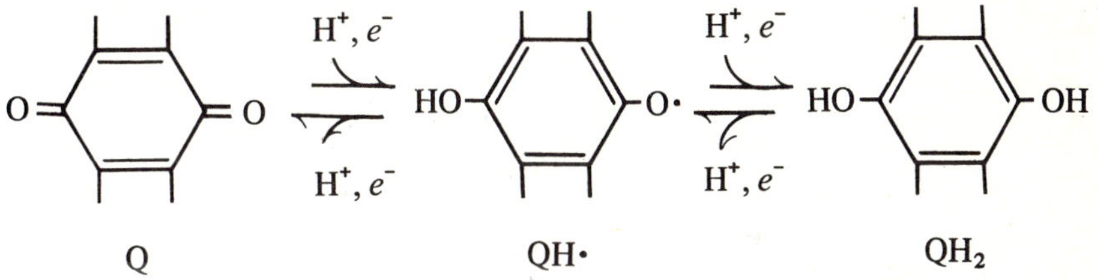

So two electrons and two protons from the matrix combine with two quinones Q to form two semiquinones QH·. Another two protons from the matrix and two electrons donated by the membrane protein cytochrome b combine to form two hydroquinones QH_2. These two QH_2 give up two H^+ to the outside (intracristal) medium, producing two QH· and two electrons. The two QH· in turn give up two protons to the outside medium and two electrons that are bound by cytochrome b, which had supplied electrons to the reaction QH· → Q. The two electrons from the reaction QH_2 → QH· are accepted by yet another intrinsic protein, cytochrome c_1, and then in sequence by cytochrome c and cytochrome a. These last three membrane proteins appear to lie at the intracristal side of the membrane. From cytochrome a the electrons go to cytochrome a_3, a protein that seems to extend to the matrix side of the membrane. Here occurs a last interaction of membrane plus electrons with matrix when the two electrons from cytochrome a_3 combine with O_2 and H^+ from the matrix as follows:

$$\tfrac{1}{2}O_2 + 2H^+ + 2e^- \rightarrow H_2O$$

Altogether six H^+ have been moved from inside the membrane to outside, and two electrons have been combined with two more H^+ and O_2 from the matrix to form H_2O. The inner membrane has transformed the energy differences between NADH + H^+ and NAD and between $\frac{1}{2}O_2 + 2H^+$ and H_2O into the free energy of the transmembrane proton gradient. (See Figure 4-7.)

The difference in proton concentration (pH) across the inner membrane constitutes an electrochemical gradient (in the spirit of Chapter 3). Consequently there is a driving force for H^+ in the concentration gradient, and there is a driving force for any ion in the electric potential gradient. It is observed that the movement of protons across the membrane from higher to lower concentration is associated with the production of ATP from ADP and phosphate ion. Two membrane proteins are essential to this association. One seems to be embedded in the membrane and is able to mediate a flow of protons across the membrane. This is called the F_0 protein. The other is closely associated with F_0, and it can be visualized, after special treatment, by electron microscopy as a knob protruding into the matrix at the inner face of the membrane. This is called F_1. F_1 can be isolated and is seen to have an ATPase activity. When coupled with F_0 in a membrane and in contact with a medium at lower H^+ concentration (higher pH) than exists on the opposite side of the membrane, F_1 and the flow of H^+ via F_0 produce one ATP molecule from ADP for each pair of H^+ passing down the gradient from higher to lower electrochemical potential. Figure 4-8 represents this mechanism. This formation of ATP from the proton gradient produced by *respiration* completes the process called *oxidative phosphorylation.*

What the mitochondrion achieves is, first, a storage of energy in a proton gradient and, second, a transformation of energy into ATP, which is accomplished by protons moving down the gradient. This is reminiscent of the engineering tactic of storing energy as water high in the gravitational field (behind a dam or pumped to an elevated reservoir) for later conversion to electric energy.

The electric potential gradient associated with proton transport powered by respiration constitutes a driving force for various substrates. An exchange of ATP and ADP across the membrane is favored, because ATP has a time-average charge of four negative electronic charges, while ADP has three. In addition, uptake of Ca^{2+} by mitochondria seems to depend on the electric potential gradient.

There is also a protein of the inner membrane responsible for an exchange of ATP and ADP across the membrane. In conjunction with the reaction AMP + ATP $\rightarrow$ 2ADP, catalyzed by adenylate kinase, one ATP crosses from the matrix to the intracristal space as one ADP crosses in the opposite direction. The adenylate kinase is in the intracristal space, so the transport of ATP by this mechanism results in no gain in ATP concentration there. But the coupled transport of ADP into the matrix provides ADP that will serve as substrate for the production of ATP coupled to the proton gradient.

The inner membrane of the mitochondrion is perhaps the quintessential example of a system that combines enzymatic reactions, membrane positioning of enzymes and transport proteins, and membrane permeability properties in an effective and stable multifunction device.

4-4 COMPARTMENTATION AND TRANSPORT

So far several basic features of biologic systems have been shown to be coupled with the existence and nature of biomembranes. Most fundamentally and simply, biomembranes introduce controlled flows and differences in concentrations from point to point in biologic systems. That is, there is compartmentation, and there is intercompartmental transport. These properties, discussed separately in Chapter 3 and in Section 4-1, were brought together in Section 4-3 in the discussion of oxidative phosphorylation. Another phenomenon that is effectively interpreted in this way is transepithelial transport.

Epithelia are described in more detail in Section 7-4. The significant structural feature of an epithelium for this discussion is that it is a layer (or layers) of cells that abut at cell surfaces within the layer and that are exposed to different extracellular compartments at the two faces of the layer. In the simplest conceptualization, then, the system presents three compartments separated by two (collectively) plasma membranes, as represented in Figure 4-9. The transcellular route for water and solutes involves the crossing of at least two membranes. (Possibilities for other routes are mentioned in Section 7-4.) Concentration differences across single biomembranes are generated by means discussed in Chapter 3, producing differences between, for example, intracellular and extracellular compartments. The generation and maintenance of differences across an epithelium require further elaboration.

An essential feature for net transport between compartments A and B of Figure 4-9 is an asymmetry in and/or around the epithelium about its central plane. This is a requirement similar to that for a single transporting membrane. In this case the asymmetry may result from a difference in the plasma membranes of component cells at the two distinct faces of the epithelium. Examples of such differences are given in Section 7-4. Here the purpose is to see how a combination of compartmentation and transmembrane transport can induce net transepithelial transport.

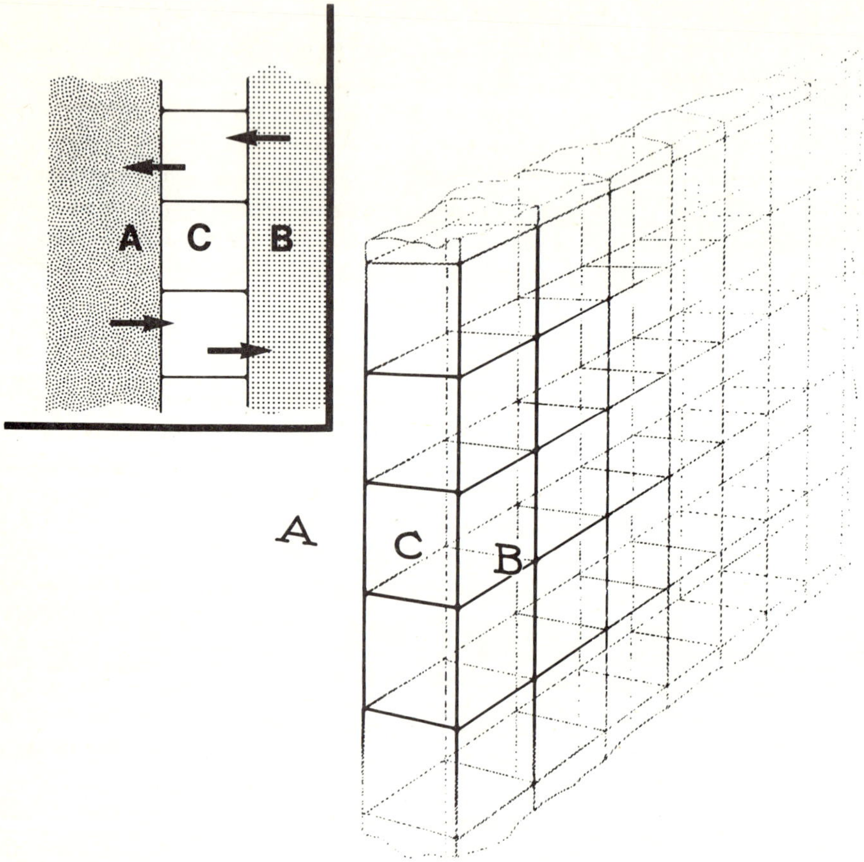

Figure 4-9 Compartment relations in a simple single-layered epithelium. The array of cells is represented on the right. The inset, upper left, shows the compartments and separating membranes: A and B, extracellular; C, cytoplasmic.

Consider the steady state for the epithelium represented in Figure 4-9. The factors for Na^+ transport, say, are the number of pumps, the rate of reaction of a pump complex, and the leak (passive permeability). Flow-coupled transport of Na^+ is ignored, and the electrostatic driving force is ignored. This simplifies the argument, while permitting contrast with the one-membrane case. Symbolically, then, the situation for unit surface area is described as follows:

$$J_{rA} + J_{rB} - P_A(c_A - c_C) - P_B(c_B - c_C) = 0$$

where the J's represent the component-reactive transport of Na^+ out of the cell (number per second), the P's represent the passive permeabilities, and the c's represent the Na^+ concentrations between and in the respective compartments. The equation is a statement of the steady-state requirement that c_C be constant. It is clear that there are many possibilities for relations among the pumps and leaks in which $c_A \neq c_B$. The analysis is complex, but it has been shown that the quantitative characterization of single biomembranes discussed in Chapter 3, when applied to a simple two-membrane system, is consistent with a transepithelial concentration difference.

In the case of circulating extracellular fluids the concentration of a solute in the reservoirs on the two sides of an epithelium may be effectively constant. Even when equal, there may be a net transport of solute across the epithelium with a constant intracellular concentration. In such a case the flow may be driven by the pumps (component-reactive transport). On the other hand, circulation of fluids—as occurs in capillaries and kidney tubules—may drive net solute transport by producing a transepithelial concentration difference.

In view of the fact that a simpler single biomembrane can produce solute and solvent flows and concentration differences, one may wonder at the evolution of the more complex epithelium. However, the membranes must be generated and, as discussed in Chapter 5, biomembranes are made by—as components of—cells. Furthermore, the glycocalyx and the anti-glycocalyx sides of the membrane are never both exposed to extracellular media. Also, the system of membranes (the epithelium), unlike a hypothetical single noncellular membrane, has a built-in apparatus capable of energy production.

4-5 SUMMARY

Biomembranes are barriers to organelles, complexes of macromolecules, and macromolecules per se. In addition, biomembranes are selective barriers to soluble substrates. These barrier properties provide protection of, and from, organelle contents and make possible storage and pulse release of cell products. Functionally related cell components are associated within organelles and on biomembranes. Concentrations of enzymes and substrates are controlled specifically in association with different membranes and organelles.

Membrane-bound enzymes allow coupling of sequential reactions and control of enzyme activity. The vectorial nature of membrane and membrane-bound enzymes brings about differentiation of aqueous regions within the cell and between cytoplasm and extracellular medium. In the case of the mitochondrion the vectorial inner membrane couples energy-yielding respiration with a transmembrane proton gradient that produces ATP in the process of oxidative phosphorylation.

The effects of compartmentation and the transport properties of biomembranes combine in structures such as epithelia to produce net transport of solutes and solvents between extracellular compartments.

The inhomogeneity maintained by membranes in cells is a prerequisite of all functions of living cells.

Chapter 5

Membrane Biogenesis and Turnover

The picture presented of the molecular organization of the biomembrane and the physical forces that account for its stability raise some interesting questions with respect to the origin of the biomembrane. Given that the amphiphilic nature of membrane components and the aqueous environment may be expected to generate the micellelike biomembrane, how is it that these membranes are so orderly and complex? They are very specifically asymmetric with respect to the plane of the membrane; they have nonrandom organization in the two-dimensional extended space of the membrane; they seem to undergo specific morphogenic sequences of change in contour and differentiation with respect to their localized structure. These problems are approached in this chapter by considering first comparisons among biomembranes and then the natural synthesis and organization–the biogenesis–of biomembranes of eucaryotic cells. The turnover of membrane constituents will be regarded here as the other half of the closed cycle that results in a dynamic balance between organelle formation and breakdown. Throughout, it will be helpful to recall that the living cell is a dynamic system in which thermal energy is a prime mover and fluctuations are important. Furthermore, it is a system appreciably far from equilibrium and only occasionally at steady state.

5-1 MORPHOLOGICAL AND CHEMICAL DIFFERENCES AMONG MEMBRANES

The morphological distinctions among organelles include differences among the membranes per se as well as differences in location, shape, and structural complexity. The morphological differences seen are, of course, differences developed by the treatment of the structure in preparation for electron microscopy. The result of fixing and staining a plasma membrane with OsO_4 is a clearly dark/light/dark (electron-opaque/electron-transparent/electron-opaque) image (positive print) in cross section. Similar treatment and examination of other membranes of eucaryotic cells reveal a less sharp (with less contrast from staining) lamination and thinner membrane. The greatest difference in this respect is between the nuclear membrane and the plasma membrane. The membrane of the endoplasmic reticulum is more like the nuclear membrane than the plasma membrane, whereas the membrane of secretory vesicles is more like the plasma membrane than the nuclear membrane. The membrane of the Golgi apparatus often appears midway between these extremes (Figure 5-1). In some cases the Golgi apparatus exhibits a transition within itself such that the membranes of the cisternae nearer the nucleus are more "nucleolemmal" and those farther from the nucleus more "plasmalemmal."

Biochemical observations support the contention that there is a graded variation among the membranes of eucaryotic cells. One of the more thoroughly examined species is the rat. The prominent cell types used have been the liver cell and the mammary gland cell and its milk fat globule membrane, which is derived from plasma membrane and (probably) secretory vesicle membrane (see Section 5-2). The biochemical markers for membranes are proteins and lipids. Protein markers may be classes defined by electrophoresis or some other physical technique, by enzyme activity, by antigenic identity, or by other more or less specific labeling. As yet there is little comparative information on proteins in membranes of different organelles, although a number of enzymes contained within organelles have been located specifically. Some membrane-bound enzymes are associated with particular organelles and are referred to as *marker enzymes*. Some of these enzymes show the relative activity level associated with Golgi apparatus to be intermediate between the levels associated with endoplasmic reticulum and plasma membrane. Electrophoretic patterns of membrane proteins differ among organelles, but it cannot be said that the patterns indicate a transition sequence.

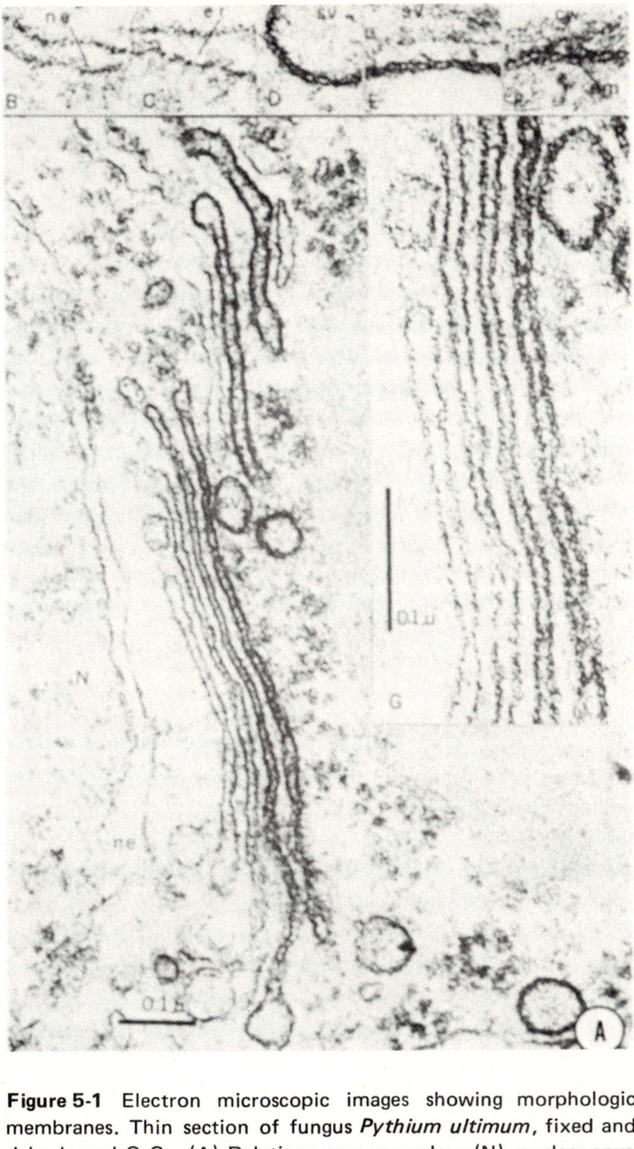

Figure 5-1 Electron microscopic images showing morphological differences among membranes. Thin section of fungus ***Pythium ultimum***, fixed and stained with glutaraldehyde and OsO_4. (A) Relations among nucleus (N), nuclear envelope (ne), dictyosome, and secretory vesicles (sv). (G) Higher magnification of Golgi apparatus and secretory vesicle; note the graded change in membranes from nuclear face to outer face of the stack of cisternae. Insets (B) through (F) show details of the various membranes; er, endoplasmic reticulum; pm, plasma membrane; sw, cell wall. (From S. N. Grove, C. E. Bracker, and D. J. Morré, *Science* 161: 171–173, July 12, 1968. Copyright 1968 by the American Association for the Advancement of Science.)

On the other hand, lipid composition varies in an interesting way. Cholesterol is generally considered a characteristic component of the animal plasma membrane, where it is in greatest proportion. It is in least proportion in the nuclear membrane; it has not been found at all in nuclear membranes of some cells and in very small proportions in others. The phosphoglycerides of the endoplasmic reticulum, Golgi apparatus, and plasma membrane vary (in the rat liver cell) in such a way that the proportion (fraction of total lipid phosphorus) in Golgi membrane is intermediate between that in endoplasmic reticulum and that in plasma membrane for sphingomyelin, choline phosphoglyceride, ethanolamine phosphoglyceride, and inositol phosphoglyceride. For example, sphingomyelin (as fraction of phosphorus) is some three times greater in Golgi apparatus and over four times greater in plasma membrane than in endoplasmic reticulum; phosphatidylcholine is decreased in Golgi apparatus compared with endoplasmic reticulum.

The chemical differences described above apply to various mammalian cells that possess a full complement of organelles, synthesize protein, and are metabolically active. The plasma membranes of these cells may have different enzymes and different carbohydrate groups in the glycocalyx, but they conform well to the characterization of the plasma membrane developed in Chapter 2. There are, however, exceptions, such as (and these are functionally opposite extremes) the outer-segment membrane of the retinal rod and the Schwann cell.

The retinal rod is a specialized nerve cell; the outer segment is largely a plasma membrane containing disks formed by invagination and separation of plasma membrane. This membrane is half or more protein by mass of which over 80% is rhodopsin, a photosensitive glycoprotein responsible for generating action potentials in response to absorbed photons. The Schwann cell is a specialized cell that wraps its plasma membrane several times around the axon in a myelinated nerve. These layers of membrane have no cytoplasm left between them; they form a laminated insulation around the axon. The plasma membrane of this myelin sheath is 20-22% protein by mass. These two are extremes of specialization that contrast with most other plasma membranes. Little information about the relations among membranes of these two cells is available except that labeled membrane components do exchange in the rod cell until they are contained in the membranes of the disks. Here they remain until the disks are extruded or destroyed at the peripheral end of the cell.

5-2 MEMBRANE FLOW

The morphological and biochemical results discussed above are very suggestive, but they do not demonstrate flow. A way of observing transfer from one organelle membrane to another that does probably reflect flow is the pulse-labeling method. A pulse of some labeled (usually radioactive) metabolite is added to the medium of a cell culture and left for a short time, after which the cells are washed and incubated further. Samples are then taken at known times; the organelles are separated, the respective membranes isolated, and the amount (or specific activity) of label determined. A similar experimental approach involves the use of tritium-labeled metabolites and autoradiography (by which the location of tritium is observed) in conjunction with microscopy to locate the label at varying times after injection into living animals.

Figure 5-2 is an idealization of the sorts of results obtained in some systems. The time scale shown may be 10 min to several hours. For example, when young rats are injected with radioactive fucose (a hexose found in membrane glycoprotein) and sacrificed at varying times after the injection, autoradiography shows the label first in the Golgi apparatus. Then it appears in secretory vesicles, increasing as the amount in the Golgi apparatus decreases. Subsequently the label appears in plasma membrane as it disappears from both the Golgi apparatus and vesicles. In other systems with other labels, the endoplasmic reticulum has been seen to be a precursor to Golgi apparatus in this apparent flow of label.

These three sorts of evidence—morphological, biochemical, and cytochemical—together constitute a strong case for the thesis that the various

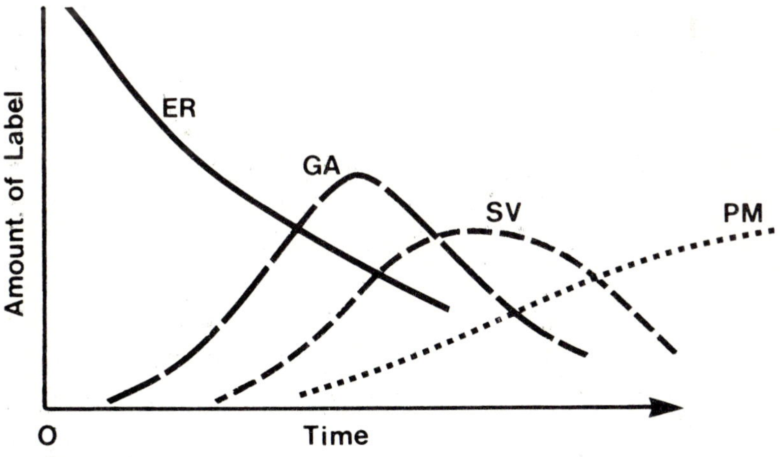

Figure 5-2 Representation of an idealized pulse-label experiment demonstrating membrane flow. The curves represent relative amounts of label in the four different membranes: ER, endoplasmic reticulum; GA, Golgi apparatus; SV, secretory vesicles; and PM, plasma membrane.

membranes of the eucaryotic cell are synthesized through a common dynamic chain, which is represented by the major path in Figure 5-3. This situation is suggested by the occasional observations of continuity of membrane between and among organelles. The possibilities of artifact or misinterpretation most assuredly exist, but the images obtained by electron microscopy also exist—and these images show continuity of outer nuclear membrane with endoplasmic reticulum, endoplasmic reticulum with dictyosome, dictyosome with secretory vesicle, secretory vesicle with plasma membrane, plasma membrane with endoplasmic reticulum, and perhaps other connections, pairwise and beyond.

The morphological evidence is subject to question, although altogether it seems coherent. The biochemical and cytochemical evidence for an assembly-line process of membrane biogenesis, however, is compelling. Apparently synthesis of membrane begins in the endoplasmic reticulum—possibly in the nuclear membrane—and proceeds as the membrane is passed on through Golgi apparatus to secretory vesicles, which pinch off the Golgi cisternae and move to the plasma membrane to which they are added by fusion.

A closer look at the Golgi apparatus (Figure 5-1) shows these two transitions. In some cases, between what is clearly endoplasmic reticulum and what is Golgi apparatus, the membrane is seen in the form of vesicles. These are sometimes called *transition elements*. There is evidence that what appears in thin-section electron microscopy as separate cisternae of Golgi apparatus and transition elements may in fact be a single, connected system. In several different cell types, including the pancreatic acinar cell, high-voltage electron microscopy of sections several micrometers thick shows a network of fine tubules connecting the cisternae at the end of the Golgi stack near the nucleus. The occasional observation of continuity between membrane of endoplasmic reticulum and Golgi apparatus was mentioned above.

During this process of membrane biogenesis, components of membrane are added and modified in different organelles. For integral proteins, evidence from various mammalian cells—intestinal mucosal cells, thyroid glandular cells, pancreatic acinar cells, and liver parenchymal cells—suggests the following sequence (although variations among systems may be expected). (See Figure 5-3.) The peptide chains of membrane proteins are synthesized on the rough endoplasmic reticulum at the sites of ribosomes bound to the membrane. Perhaps one of the carbohydrate groups of glycoproteins is added specifically to peptide chains in endoplasmic reticulum, or possibly some of the longer carbohydrate chains form here before conjugation to protein. Most of the addition of carbohydrate to glycoprotein occurs in the cisternae of the dictyosome. The orientation of the integral glycoproteins is the amino end of the peptide chain at the inside face (cisternal

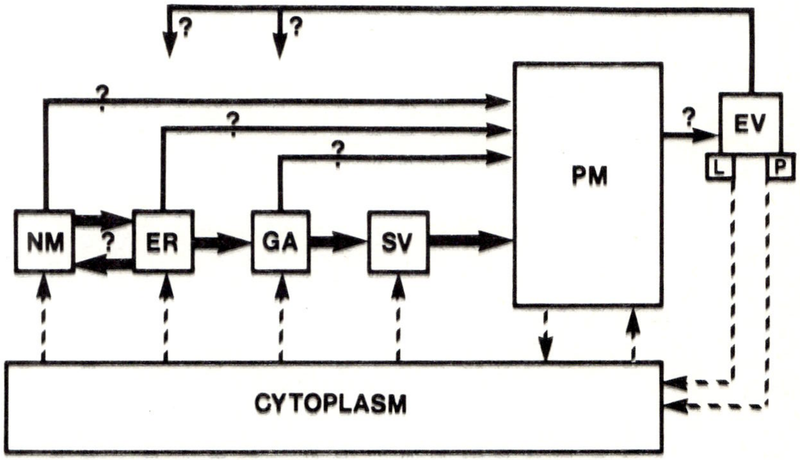

dissolved-precursor pathway

pathway with weaker supportive evidence

PM	plasma membrane	GA	Golgi apparatus	L	lysosome
NM	nuclear membrane	SV	secretory vesicle	P	peroxisome
ER	endoplasmic reticulum	EV	endocytic vesicle		

Figure 5-3 Precursor relations among biomembranes. The various paths are suggested by morphological, compositional, and pulse-label evidence. The major path from ER to PM is well supported and probably quite general among cell types.

content side, opposite the cytoplasm) of the membrane. The addition of monosaccharides is an exchange reaction from a carrier molecule to the glycoprotein, catalyzed by specific glycosyltransferases found most often in association with the contents of the Golgi apparatus.

These glycosyltransferases are found, in most cells studied, predominantly in the Golgi apparatus, and they are largely membrane-bound. The carrier molecules for the saccharide substrate in the transfer reaction are formed by a phosphate or pyrophosphate linkage of the monosaccharide to a nucleoside or to a long-chain unsaturated alcohol (dolichol). This is interestingly similar to the nature and role of tRNA (Section 5-3). The transfer reaction results in the formation of an *O*-glycosidic bond to a sugar of the glycoprotein or to the peptide chain by a hydroxyl group of an amino acid or in the formation of an *N*-glycosidic bond to asparagine or to *N*-acetylglucosamine.

Still less is known of the synthesis and assembly of lipids in the various organelle membranes. Lipid synthesis on the biochemical level has been studied quite successfully, and much is known about it. But the place where, in the assembly line described here, the particular lipids are added to membrane is unresolved. Still, the biochemical compositions do suggest a transition from endoplasmic reticulum membrane to plasma membrane.

The high fraction of lecithin in the endoplasmic reticulum of rat liver cells is strong evidence that this phospholipid is incorporated there (or possibly in nuclear membrane) rather than in dictyosome or later in the putative membrane-flow stream. Any concentration-driven diffusion in the two-dimensional membrane space would be from endoplasmic reticulum to Golgi apparatus if the overall membranes are static.

If the lipid components are synthesized in the contained media of endoplasmic reticulum and dictyosome, they may well undergo transition to a lower free energy state by entering the surrounding membrane with its adjacent hydrophobic and hydrophilic regions. Then, if membrane flow is sufficiently faster than simple (two-dimensional) diffusion within the membranes—fast enough to prevent diffusional equilibrium—the distribution of the various lipid constituents is consistent with the postulate that specific synthesis occurs in specific cisternae of the continuous system.

Out of all this evidence from a variety of systems and techniques—some quite reliable, some merely suggestive—a coherent picture emerges. The organelles of a living eucaryotic cell seem to be in part interconnected pools of membrane in a dynamic, generally steady-state system. If this is so, another question arises immediately. What is happening to prevent continuing growth of plasma membrane? There must be loss of material. Is this loss occurring molecule by molecule? To the inside? To the outside? Or is there removal of intact pieces of membrane?

Probably the major steady-state losses of plasma membrane are to the interior of the cell. The mechanism is not known, but an interesting possibility is that endocytic vesicles form from invagination of plasma membrane. This is represented tentatively in Figure 5-3. Such vesicles might fuse with lysosomes and/or peroxisomes, resulting in enzymatic breakdown of membrane and return of the products to the cell's pool of metabolites. Here they are available for synthesis of membrane, thus completing the cycle. A more efficient return would be a mechanism for recycling whole pieces of plasma membrane by endocytosis (Chapter 6) and return to the biogenic stream *in toto*. Such a process would be consistent with the observations that specific parts of plasma membrane—formerly secretory vesicle membrane—become membrane of endocytic vesicles (observed in nerve endings and anterior pituitary cells) and return to the dictyosome (observed in anterior pituitary cells).

The chain of events described so far bypasses the remaining three cytoplasmic organelles: lysosome, peroxisome, and mitochondrion. Although there is little evidence to support the idea, it is inviting to imagine a branching pathway by which lysosomes and peroxisomes originate in the way secretory vesicles do. Two qualifications are necessary. Firstly, a mechanism would be required to specialize a particular part (possibly at a particular time) of the dictyosome and/or endoplasmic reticulum so that the appropriate contents for a lysosome or peroxisome are synthesized and packaged. Secondly, the resulting vesicle must differ from the secretory vesicle in such a way that fusion with plasma membrane does not occur. In fact, apparent continuity between Golgi cisternae and lysosomes has been shown by electron microscopy.

Also, there is some evidence of specialization of regions of the Golgi apparatus. In some mammalian cells, such as rat liver parenchymal cells, morphological differences have been reported among the different saccules comprising the dictysome. Cytochemical differences have also been reported. More convincing is the specificity of proteins associated with different classes of secretory vesicles and lysosomes. The membrane glycoproteins of zymogen granules of the pancreatic acinar cell apical membrane (Section 5-3) are similar to those of plasma membrane, whereas in the lysosome membrane and mitochondrial outer membrane these glycoproteins are absent or very nearly so. All this evidence is consistent with the hypothesis that the path of biogenesis of membranes branches in the dictyosome, which is composed of regions specialized for the generation of particular membranes.

The mitochondrion is a very special case. Its possible unusual evolutionary origin was mentioned in Chapter 1. It contains DNA and the various RNAs required for synthesis of at least some of its membrane proteins (perhaps 10) associated with the inner membrane. Much has been learned of the mitochondrial genetic control of the synthesis of inner membrane proteins, with much of the information coming from yeast and fungal systems. But the actual assembly of the elements of mitochondrial membranes and the mechanism by which the two membranes arrive at their final relationship is not known. It is an attractive idea that the inner membrane is organized largely from components synthesized under the guidance of mitochondrial DNA plus components gleaned from the cytosol at rates determined by cytoplasmic conditions (in turn determined by nuclear genetic and extracellular conditions) and that the outer membrane is the result of a branch in the membrane biogenesis described above that encapsulates the inner-membrane sac. Such a process is suggested by the few published electron microscopic pictures seeming to show continuity between endoplasmic reticulum membrane and mitochondrial outer membrane. For now, however, such a picture must be regarded as anticipation of, and guidance for, work yet to be reported.

As the evidence for connectivity among membranes has increased, the term *endomembrane system* has been applied by a number of authors to this complex assembly, particularly to the endoplasmic reticulum-dictyosome-vesicle components.

5-3 SECRETION

A variety of cells in mammals are specialized to secrete particular substances, including hormones, neurotransmitters, mucus, and enzymes. In the cases of small molecules and ions, one might look for processes such as those described in Chapter 3. However, what is normally referred to as secretion in the context of cell and organ physiology is a more elaborate process very closely related to membrane biogenesis.

Much of the information in Section 5-2 is derived from investigations that began as inquiries into protein synthesis and secretion. The connection between cell organelles and protein secretion was suggested by observations made a century ago, which related the cytological appearance of pancreatic acinar cells to the catalytically active pancreatic secretion. The mammalian pancreas has been a major subject for application of the very effective separation and ultramicroscopic techniques developed in the last two or three decades. The major morphological features of the pancreatic acinar cell are shown in Figure 5-4. The acinar cell is a unit of a monolayer of like cells that form a sac around the lumen, which is continuous with the lumen of the system of ducts that carry the pancreatic exocrine secretion to the intestinal lumen.

The pancreas can be stimulated to secrete in a variety of ways.

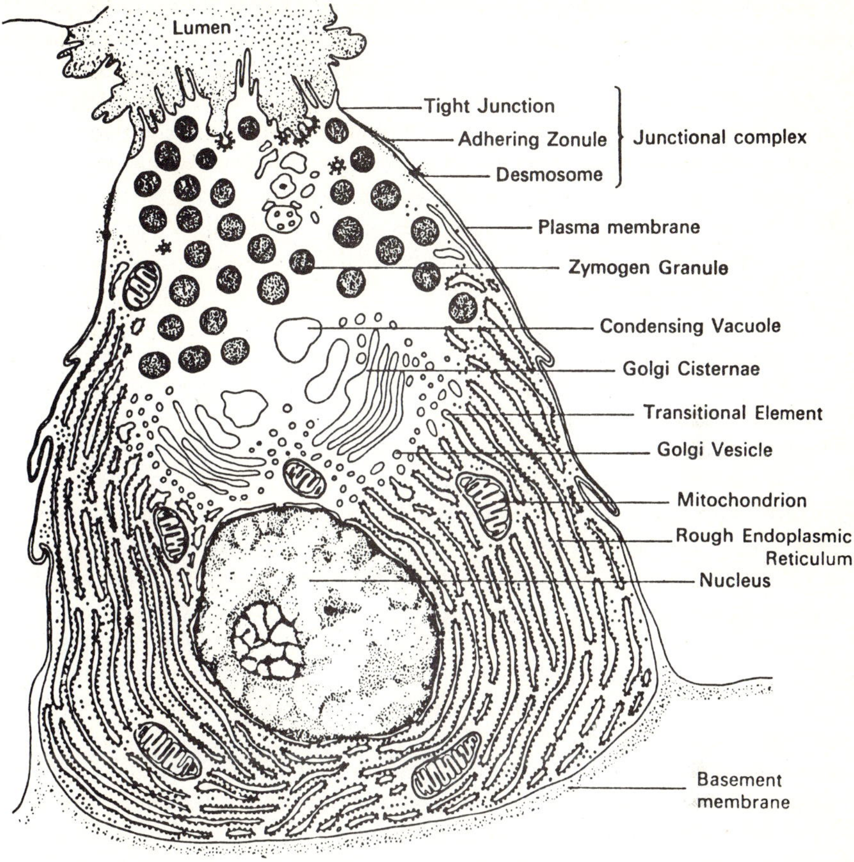

Figure 5-4 Drawing of pancreatic acinar cell. The lumen is continuous with the pancreatic duct along the path of secretion into the intestine. (From R. M. Case, Synthesis, intracellular transport and discharge of exportable proteins in the pancreatic acinar cell and other cells, *Biol. Rev.* 53: 211, 1978, Cambridge University Press.)

Physiologically it is stimulated by hormones from the intestinal mucosa and by nerves. The use of physiological and other stimuli makes it relatively simple to show a correlation between the appearance of pancreatic secretion and the disappearance of zymogen granules of the acinar cells. The zymogen granules, separated from other cell components, are found to be biomembrane vesicles whose contents correspond to the protein content of pancreatic exocrine secretion. Also, they contain the inorganic ions that characterize the pancreatic secretion. Furthermore, in some species, the plasma membrane adjacent to the lumen (apical plasma membrane) is seen to increase in area when secretion occurs. Thus, the final stage of secretion in the pancreatic acinar cell seems to be a process of fusion of zymogen granule (secretory vesicle) membrane with apical cell membrane (plasma membrane)—similar to the final stage of the biogenesis of plasma membrane as described in Section 5-2.

Autoradiography at both light and electron microscopic resolution and pulse-labeling studies have been used to corroborate this picture of the final stage of secretion and also to uncover the earlier stages. The essential difference between these investigations and those mentioned in connection with membrane biogenesis is that here the investigators are following the secreted proteins rather than membrane proteins or other membrane constituents. An experimental advantage is that the secretory products are generally produced in larger amounts, and production can be manipulated by specific stimuli. For secretory cells—those in the exocrine pancreas, mammary gland, thyroid gland, islets of Langerhans (insulin-secreting), liver, parathyroid gland, anterior pituitary gland, prostate gland, and others—without stimulation the maximum labeling of the Golgi apparatus occurred at some tens of minutes after the pulse and of the secretory vesicles at one to several hours. In studies of pancreatic acinar cells pulse-labeled with a radioactive amino acid and followed by both separation of cell components (by using density gradient centrifugation) and autoradiography, a time sequence similar to that portrayed in Figure 5-2 was found: endoplasmic reticulum was maximum at about 3 min, Golgi at about 10 min, condensing vacuoles maximum with a slight labeling of zymogen granules at 40 min, and zymogen granules maximum after 1 or 2 h.

Thus far the visible beginning of the process of synthesis and secretion in the pancreatic acinar cell is the appearance and activation of mRNA carrying the code that determines the amino acid sequence of each of the several proteins. The mRNA induces two fragments from the cytoplasm to bind to the MRNA strands to form ribosomes. The ribosome-mRNA complex interacts with the membrane of endoplasmic reticulum. Probably this interaction produces a channel through the membrane, because the peptide chain synthesized by these ribosomes appears within the endoplasmic reticulum cisterna. The first part of the amino acid sequence corresponds to the nucleotide base sequence of the end of the mRNA strand that initiates the translation; within the endoplasmic reticulum it is removed from the major portion of the peptide that constitutes the secretory protein.

This mechanism of simultaneously synthesizing a peptide chain and introducing it into the cisterna of the endoplasmic reticulum is probably similar to the process of synthesizing integral membrane proteins, except that the latter may well never be free in the cisterna but rather remain embedded, at least partly, in the membrane. In either case it is established that the peptide chain (at least its various labeled constituents) subsequently appears in the dictyosome. In the case of membrane proteins this is consistent with what was called membrane flow in the preceding section. In the case of secretory proteins the question is whether packages (membrane plus contents to be secreted) are passed along or whether just contents flow through the system. That the secretory products are stored in the zymogen granules and that fusion of zymogen granule membrane with apical plasma membrane accompanies secretion from the cell suggest that the flow of secretory products must be accompanied by a flow of membrane. The evidence points fairly clearly to the notion that secretion and membrane biogenesis—at least in specialized secretory cells—are coupled processes.

Exocrine pancreatic secretion contains inorganic constituents in high concentrations as well as specific proteins. Some of these high levels [of Cl^- and bicarbonate ion (HCO_3^-), for example] may result from specific transmembrane transport features of plasma membrane of the acinar cell, but others seem related to the activity of intracellular organelles. There is evidence that Ca^{2+} is concentrated in (or on) the dictyosome in various secretory cells. In the pancreatic acinar cell Ca^{2+}, Mg^{2+}, and Zn^{2+} are concentrated in zymogen granules. Zn^{2+} uptake parallels amino acid uptake in double-label experiments when secretion is stimulated. On the other hand, Ca^{2+} appears in pancreatic secretion before newly synthesized protein, and Mg^{2+} appears as Ca^{2+} appears.

The concentration of inorganic ions in zymogen granules raises a question also suggested by the extreme opacity to electrons of zymogen granules, as demonstrated by electron microscopy. If the zymogen granules are indeed formed from condensing vacuoles, as the evidence suggests, then the concentration of protein and divalent cations in the zymogen granules would seem to involve a loss of H_2O and therefore of volume from the condensing vacuole-zymogen granule. Such a situation leaves two problems to resolve. One is the mechanism by which the membrane area is adjusted to the altered volume. The zymogen granules are generally spherical, whereas the condensing vacuoles are more often irregular in shape. The sphere is the form of maximum volume-to-surface ratio. Even if the condensing vacuole were spherical, a loss of half its volume in transition to a spherical zymogen granule would require a loss of 37% of its membrane (assuming smooth surfaces). Geometrically it is impossible to preserve both volume and surface area by fusing or dividing spheres. This aspect of secretion is an unresolved problem.

The other problem raised by the condensation part of zymogen granule formation is that of the required transmembrane transport conditions. If the process results from membrane transport, it must require considerable energy. However, metabolic inhibitors do not prevent passage of autoradiographically located label from condensing vacuoles to zymogen granules Furthermore, isolated zymogen granules are resistant to osmotic shock, even to suspension in pure H_2O. These observations seem more consistent with the sort of process discussed in Section 3-3 than with a process involving only membrane. The argument applies to the granules of a variety of secretory cells. Some studies of the binding of divalent cations to protein and of interactions among secretory protein molecules suggest the possibility that molecular aggregation explains condensation and reduction of osmotically active concentrations in secretory granules. Also, amphiphilic lipoproteins have been found associated with secretory granules, which suggests the possibility of micelle formation. The answer is not yet in hand, but the secretory granules may well be the best example to date of the mechanism proposed in sorption theory—that is, a cooperative physiochemical interaction among membrane-contained proteins and between contained proteins and ions. If such a process occurs in newly formed condensing vacuoles, the binding of ions and polymerization of protein molecules would produce a dilute solution phase and a passive outflow of H_2O.

5-4 SUMMARY

The differences in morphology and in protein and lipid composition of the various classes of biomembranes of the eucaryotic cell are coupled with the biogenesis of membranes. There are graded variations in ultramicroscopic appearance, in relative proportions of lipids, and in the nature of the integral proteins, which suggest a transition from the character of nuclear membranes and endoplasmic reticulum to plasma membrane (dictyosome and secretory vesicles being intermediate). Results of pulse-labeling experiments are consistent with the concept of ordered synthesis and flow of membrane from nucleus to periphery, resulting in progressive modification of the various membranes. Lysosomes and peroxisomes may derive membrane from specialized branches in this flow at the Golgi stage.

Synthesis of secretory proteins and the storage and secretion thereof are coupled with membrane biogenesis and mediated by membrane interactions with ribosomes at the endoplasmic reticulum and with plasma membrane at the cell periphery.

Membrane and/or its constituents are returned to the cycle of membrane biogenesis. In at least some cases this return involves endocytosis and fusion with lysosomes and probably peroxisomes.

Chapter 6

Biomembrane Fission and Fusion

Unlike other physical entities, biologic things are generated only from like predecessors. Planets may develop from interstellar dust; lightening may derive from wind and water; and crystals may grow from a solution. However, an organism comes only from one or more like organisms and a cell from one or more like cells. On the molecular level, DNA comes only from like DNA. Between the molecular and cellular levels of animal organization (see Figure 1-1)–at the supramolecular level–a similar generalization is quite probably true: a biomembrane is derived only from one or more like biomembranes. The ultimate step in generation of one membrane from another is either fission of a biomembrane or fusion of two biomembranes.

That fission and fusion of biomembranes are essential steps in normal cell functions was illustrated in Chapter 5. Membrane biogenesis and secretion seem to include both the separation of a single continuous membrane into two separate membranes (fission) and the joining of two separate membranes to form a single continuous membrane (fusion). Other long-familiar biologic phenomena in which biomembrane fission and fusion play a major part include mitosis and phagocytosis.

In this chapter a generalization about biomembrane transformation is presented along with some terminology. Then the fission-fusion process is discussed in the spirit of Chapter 2, followed by brief descriptions of some biologic processes involving fission and fusion.

6-1 CONSERVATION OF BIOMEMBRANE ORIENTATION

The incorporation by a cell of a solid particle, such as another cell, a cell fragment, or bit of carbon, was observed long ago. This process is known as *phagocytosis.* A similar process, called *pinocytosis,* is the incorporation of a bit of extracellular liquid. It has long been recognized that matter foreign to the dominant cell is enclosed in a biomembrane during these processes. The presence of the biomembrane in these processes and the observation of processes (such as secretion) that seem to be similar but are, in a sense, reversed have led to a newer terminology. One more often hears of *endocytosis* and *exocytosis.* Endocytosis includes phagocytosis and pinocytosis, which are special cases differentiated by the nature of the contents of the *endocytic vesicle* (or equivalently, by the nature of the stimulant). Exocytosis includes secretion that may produce no extracellular membrane, as in most cases (Section 5-4), or that may produce an extracellular membranous vesicle, as in milk from the mammary gland.

With the exception of the latter case, which is exclusively fission, exocytosis is a fusion of biomembranes. Endocytosis is a fission of biomembranes. In both cases the relationship between the sides of the membrane involved and the adjacent media is preserved (see Figure 6-1). This is seen in the processes of membrane biogenesis and secretion. There the integral proteins of the membranes are synthesized in the endoplasmic reticulum in such a way that the amino end of the peptide chain (when exposed) is on the side opposite the cytoplasm, and the carboxyl end is on the side adjacent to cytoplasm.

Where glycosylation occurs in the endoplasmic reticulum and/or Golgi apparatus, the glycocalyx is opposite the cytoplasm. This orientation is maintained for all organelles and stages of biogenesis. When fusion of a vesicle with plasma membrane occurs, the same side of membrane remains associated with cytoplasm. For the other side, the extracellular liquid (or the surface of an adjacent cell) takes the position of the various intracisternal media.

The process of endocytosis begins with a stimulus to the plasma membrane, which initiates the invagination of the membrane. The invagination results in a pocket the concave surface of which is the glycocalyx side of the membrane exposed to extracellular medium. The invagination proceeds

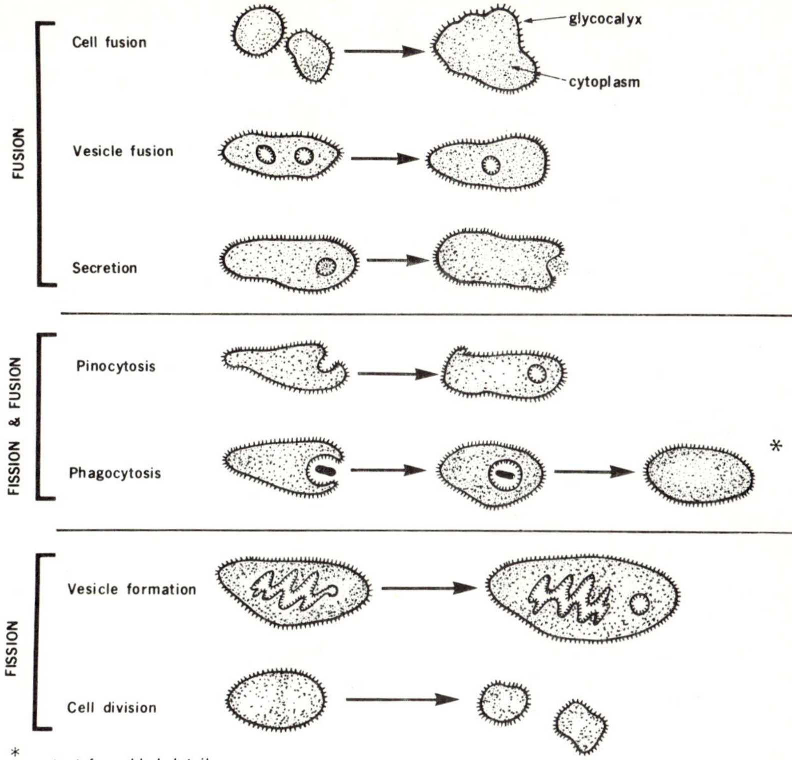

Figure 6-1 Relations of compartments and membrane orientation in fission and fusion. The orientations of the membranes with respect to cytoplasm are maintained in both fission and fusion.

until there is a small point of contact—continuity—between the plasma membrane and the new spherical vesicle; the latter then breaks off and is surrounded by cytoplasm. Again, the medium-membrane relationships are preserved.

One might honor the consistency of these relationships by formulating the law of conservation of biomembrane orientation: biomembrane transformations occur only in ways that preserve the contiguity of glycocalyx (or amino end of membrane-spanning integral peptide chains) with non-cytoplasmic media. Figure 6-1 illustrates this condition for several phenomena.

This restraint on biomembrane relations may be seen as a consequence of the mechanism of biogenesis. As such it originates in the directional reading of the genetic code by ribosomes and the means of attachment of ribosomes to endoplasmic reticulum as described in Section 5-3. If one assumes that the asymmetry of phospholipid composition of the coupled monolayers of the membrane is a consequence of the orientation and nature of the integral proteins, then the die is cast—entirely—at this point. The roles of membrane receptors in cellular interactions and responses to external stimuli such as hormones make this rule of biomembrane orientation critical to virtually every physiological process known.

An exceptional secretory process occurs in the lactating mammary gland. The milk fat is collected into globules in the cytoplasm, probably by simple physicochemical phase separation. These globules have no enclosing membrane. When fat is secreted from the cell, the globule pushes outward until it is surrounded in an evagination of plasma membrane that pinches off to produce the membrane-bounded milk fat globule of milk. A simultaneous secretion of protein and other milk components occurs by the usual exocytosis described in Section 5-3, balancing the plasma membrane loss from fat secretion. In both modes the law of conservation of biomembrane orientation holds.

An important exception to this rule has been found recently in the red cell membrane. Labeling experiments indicate that the carboxyl group end of the peptide chain of band 3, the most abundant single membrane glycoprotein (or group of indistinguishable glycoproteins), is on the outside of the membrane, and the amino group end is on the cytoplasmic side. Still, the carbohydrate groups are on the extracellular side and are a major part of the glycocalyx. Biology is notorious for breaking its own rules.

6-2 PHYSICAL CHEMISTRY OF FISSION AND FUSION

In Section 2-5 the relatively high free energy of states in which membrane edges were exposed to an aqueous medium was cited as the driving force for the resealing of torn membranes composed of bilayers of amphiphilic molecules (see Figure 2-18). Such considerations are basic to the understanding of membrane fission and fusion. One may view the two processes as transitions between states of relative stability (relatively low free energies). Within such a conceptual framework, the process of fission of a closed biomembrane sac into two closed sacs may be separated into distinct phases or states. The two end states are both stable; the hydrophobic core of the bilayer is continuous with itself and not in contact with water.

Spontaneous fusion and fission of the biomembrane sacs is prevented by the existence of at least one transition state of free energy that is higher than either end state by considerably more than thermal energy kT. (See the discussion of free energy in Section 2-1.) In biologic systems fission and fusion occur under a variety of conditions and are affected by a variety of physical and chemical factors. But whatever the situation (that is, the initiating event), the critical phase is the intermediate one. Either the system must pass over an energy barrier analogous to the activation energy of a chemical or physical transition between stable states, or the energy barrier must be sidestepped by a specific path such as that offered by a catalyst for a chemical reaction (enzyme in the biologic case).

The simplest case with any realistic similarity to the biologic one is the fusion or fission of artificial phospholipid or phospholipid-cholesterol vesicles, liposomes, suspended in an aqueous medium. The stages of the process–in either direction–are illustrated in Figure 6-2. For fusion there must be contact, a transition phase, and shape change; for fission there must be shape change, a transition phase, and separation. The contact and separation phases involve energy changes related to two kinds of forces that dominate the interaction of similar charged surfaces: electrostatic force and van der Waals (electrodynamic) force. The electrostatic interaction is complicated by the point nature of the charges, the mobility of the charges in the surfaces, and the counterions and polar water molecules in the adjacent medium. Nevertheless it is quite probably true for liposomes of natural phospholipids such as lecithin that the electrostatic force is repulsive at very close distances of separation (less than a few nanometers). The van der Waals force, on the other hand, is attractive and significant only at distances of nanometers or less. Calculations for planar structures bearing fixed charges arrayed as in lipid bilayers indicate a net energy of interaction with a shallow minimum at a separation of 4–8 nm. In any case, coming into molecular contact (that is, no water or solvents between the phospholipids of the membranes) involves an energy barrier. Separation, in contrast, may be expected to occur spontaneously in the absence of other factors.

The calculations mentioned above were for planar surfaces. The situation for membranes with separated point charges will depend strongly on the curvature of the membrane and the distribution of charges with respect to the curvature. In fact, some observations of the fusion of cells are interpreted as indicating that a small radius of curvature is required. It is claimed that cells fuse only at points where microvilli exist and that secretory vesicle fusion can occur only because such vesicles have a small radius of curvature.

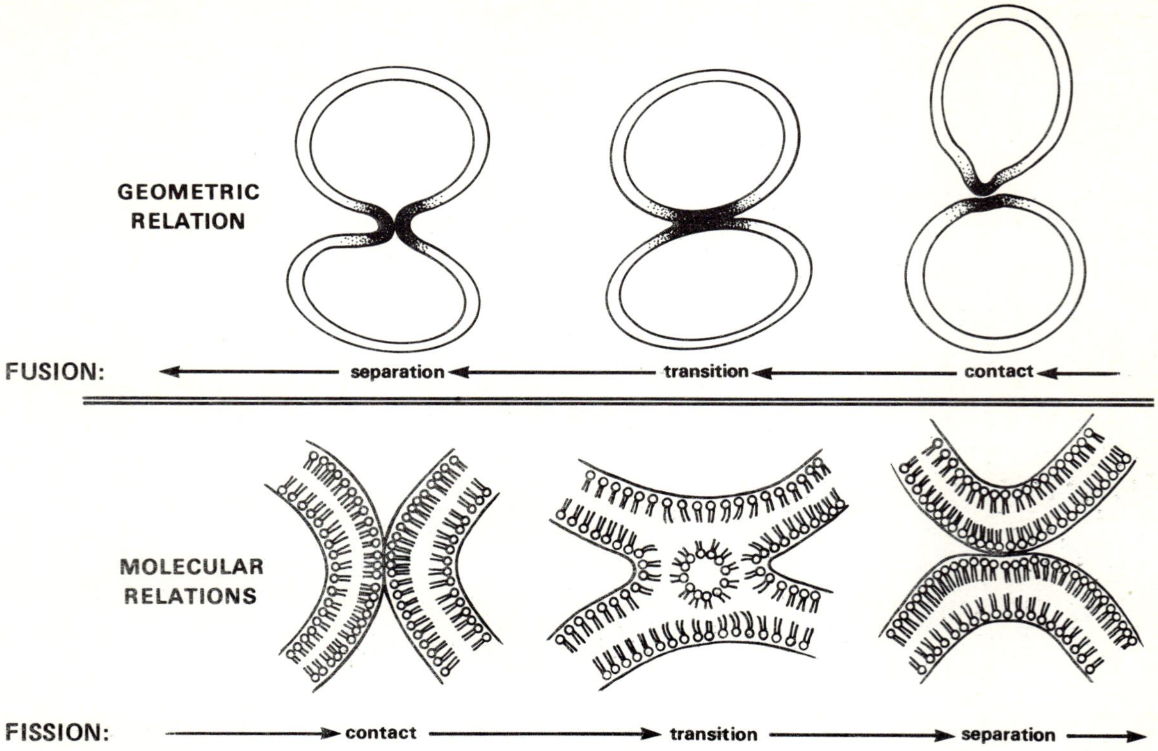

Figure 6-2 Correlation of molecular configurations with geometric relations of fusion-fission of phospholipid liposomes. The configuration of molecules shown in the transition stage is hypothetical, although some disorder or unstable configuration is necessary.

Fusion does occur between liposomes. But to occur appreciably, conditions—such as temperature and the presence of cations—must exist that will carry the system through the transition phase. Artificial liposomes that do not fuse in aqueous suspension have been shown to fuse when sufficient Ca^{2+} (2 m*M* for choline phosphoglyceride vesicles) is present in the solution. However, this is not simply charge neutralization reducing the electrostatic repulsion between liposome surfaces, for the fusion induced by Ca^{2+} occurred only when the temperature of the system was greater than the transition temperature (T_c) for the crystalline to liquid-crystalline phase change of the lipid bilayer. Other ions can induce liposome fusion. Mg^{2+} can, but again only when the temperature exceeds T_c.

In additon to cations, other substances induce fusion of liposomes. Among these are lysolecithin, which was mentioned in Section 2-1 (p. 19), and dimethylsulfoxide. The inclusion of lysolecithin in phospholipid membranes increases fluidity of the hydrophobic core; this is a result of its disruptive effect on the molecular order. Dimethylsulfoxide is an effective solvent for lipids and enters biomembranes in aqueous suspensions. It, too, quite probably relaxes the usual orderly arrangement of phospholipid bilayers.

Liposome fission occurs when liposomes are energized by ultrasound. In fact ultrasound is used as a means of preparing suspensions of liposomes of more or less uniform characteristics. Such a process might be described as a sophisticated sledgehammer forcing the system through the transition state to another stable end state (smaller vesicles) by mechanical agitation at the molecular level.

6-3 CELL FUSION

Fusion of the plasma membranes of cells can be induced in the laboratory in a variety of ways. Such experiments offer the possibility of unraveling the mechanisms of contact and of the transition phase by which the two membranes become one. However, the problem is considerably more complicated with natural biomembranes than with artificial liposomes. With natural membranes one is dealing with coupled monolayers of several distinct phospholipids, glycolipids, and cholesterol. The distributions of these lipid components are unknown. Do they exist in separate domains? Do such domains have different physical properties? In addition, there are the membrane proteins. The integral proteins, many of them glycoproteins, present receptors of a very specific nature to the external medium and to the membrane of the other cell. They (at least some of them) are mobile within the membrane, possibly polymerized, and subject to aggregation and possibly to conformational change. Peripheral membrane proteins, interacting with membrane components and cytoplasm, may well alter local membrane curvature in such a way as to affect the steps necessary for fusion.

The evidence suggests that the fusion of two cells is a complex process and that it can be initiated in essentially different ways. Fusion can be induced in the laboratory with a wide variety of agents. The erythrocyte provides a relatively simple example of a natural membrane. Avian erythrocytes are nucleated but have homogeneous cytoplasm within the plasma membrane. The nuclei serve as markers for observing cell fusion, which produces multinucleate cells. With such a system, investigators have induced fusion with dimethylsulfoxide at 37°C, with glycerol plus dextran, and to a lesser degree with glycerol alone. In all these cases, Ca^{2+} was required. Fusion has been induced by various sugars, glycerol-like compounds, substituted glycerols, and α-tocopherol (vitamin E). Some of these agents act synergistically in producing fusion. It seems that the process involves at least several steps and more than one component of the membrane. Or, alternatively, there are quite different mechanisms operating in some of the cases involving different classes of inducers of fusion.

A particularly interesting class of fusion-inducing compounds is the fatty acids. Saturated forms of these acids cause fusion of chicken erythrocytes when the chain length is 10–14 carbons; they are inactive below and above this range. Unsaturated forms are active both below and above this length. It cannot be said with certainty that this results from the behavior of fatty acids within the membrane, because chain length and saturation also affect their behavior in water and at the water-membrane interface. Some effects of double bonds on monolayer behavior were discussed in Section 2-2.

There are viruses that induce cell fusion. The Sendai virus, an influenza-type virus bearing a biomembrane, has been used with human red blood cells to explore the requirements for membrane fusion. Both whole erythrocytes and ghosts (closed membranes) produced by hypotonic medium (p. 61) can be fused by the Sendai virus. The mechanism is probably more specific than in the case of liposome fusion; the cells and the virus in this case have more or less specific receptors that determine the contact and possibly the transition phases of fusion. The natural process is a fusion of cell membrane with virus membrane so that the viral genome is injected into the cell. In these experiments the virus particles are used as inducers of fusion between erythrocytes, which is more easily detected than fusion of viral and plasma membranes. It is found that divalent cations are required. In fact, Ca^{2+} can induce fusion between human erythrocytes without virus when the suspending medium has a sufficiently basic pH (low concentration of H^+). ATP is not required for virus-induced fusion. Iodoacetate, which poisons glycolysis (the only metabolic energy source of erythrocytes), does not prevent fusion.

6-4 ENDOCYTOSIS AND EXOCYTOSIS

Much of what has been said of biomembrane fission and fusion pertains to artificial situations. Cell-to-cell fusion does occur naturally, but it is rare among mammalian cell types. On the other hand, both fission and fusion of cellular organelles are normal processes for most cells of unicellular and multicellular organisms. In these normal situations one speaks of *endocytosis* and *exocytosis.* Exocytosis was discussed in Chapter 5; membrane biogenesis and exocrine secretion are prime examples of exocytosis. Endocytosis has the appearance of being the reverse of exocytosis and, in fact, is probably the standard mechanism for maintaining plasma membrane in cells undergoing exocytosis.

In general, exocytosis and endocytosis result in a change in the number of membrane-separated compartments. (See Figure 6-1.) A single fusion event reduces the number of compartments by one. A single fission event increases the number of compartments by one. Examples of the former are fusion of a secretory granule with plasma membrane and (possibly) fusion of a transition element with dictyosome. Examples of the latter are formation of a secretory granule from dictyosome cisterna and (possibly) endocytic recycling of plasma membrane.

It is clear that both fusion of intracellular organelles and exocytosis are subject to specific controls. Intracellular fusion occurs only in certain combinations; examples will be discussed below. Exocytosis occurs in response to specific stimuli. Both sorts of control characterize the pancreatic acinar cell (Section 5-3). On the one hand, zymogen granules generally do not fuse with one another. However, occasionally acinar cells are seen that appear to have had a zymogen granule fuse with a zymogen-granule-become-plasma-membrane to form a deep invagination, as if the required change in membrane is propagated after fusion. On the other hand, exocrine pancreatic secretion responds to a variety of stimuli with the disappearance of most of the zymogen granules. Certain hormones and drugs produce this effect.

The specificity of chemical stimuli of exocytosis is not understood. It seems certain, however, that at least one stage of the process involves a specific receptor. Precedent suggests that such a receptor is likely to be a membrane component. It may be in the plasma membrane or in the vesicle membrane. The specific response may be one that produces a specific contact followed by a more or less nonspecific fusion, or it may be a specific detour around the high-energy barrier of the transition phase of fusion.

How secretory granules are brought to fuse with plasma membrane is an open question. The prevailing hypothesis is that microtubules and microfilaments play a central role, at least in the contact phase. Microtubules are protein structures in the form of hollow cylinders some 20 nm in diameter and sometimes micrometers in length. Microfilaments are proteins about 5 nm in diameter, extending over tens or hundreds of nanometers. The component proteins are similar antigenically and morphologically to actin and myosin and are widely believed to be contractile. Secretory vesicles are seen in many cells to be associated with these cytoplasmic elements in such a way that the vesicles may be confined by them and moved by them to the plasma membrane in response to stimulation of secretion. An important point here is that both Ca^{2+} and ATP (or a metabolic source of ATP) are required for exocytosis.

There is indirect evidence beyond the morphological association for the role of microtubules and microfilaments in exocytosis. There are agents that more or less specifically disrupt—dissipate or fragment—these filamentous supramolecular structures. In the case of microtubules, the natural compounds colchicine, vinblastine, and vincristine depolymerize the protein components. In the case of microfilaments, the antibiotic cytochalasins, particularly cytochalasin B, disrupt the normal morphological appearance. There is some question as to the specificity of these agents. What other cell components might they affect? Nevertheless, there is a consistency of effects that has led to qualified acceptance of these agents as probes of microtubule or microfilament involvement.

Vinblastine and colchicine inhibit enzyme secretion in the pancreas of several mammals, suggesting a role for microtubules. Similar effects have been found in adrenal medulla, parotid gland, submaxillary gland, thyroid gland, liver, and anterior pituitary gland. Much interest has been raised by evidence of similar involvement of microtubules in the rather extensive movement of vesicles containing neurotransmitters along presynaptic axons to the synaptic terminals. The implication of microfilaments by exposure of secreting cells to cytochalasin B is less clear, because exposure to the agent has been associated with inhibition of stimulated secretion at one concentration and with no effect or increased secretion at a lower concentration. Cytochalasin B does consistently cause reversible morphological changes—namely, loss of microvilli and of the complex of microfilaments at the bases of the microvilli (the *terminal web*).

The fusion of plasma membrane and secretory vesicle has been studied morphologically in a variety of cells. (Refer to Figures 6-2 and 6-3.) These investigations have indicated that the apposed membranes form a five-layered structure that becomes three layers, as with a single unit membrane, and that this region thins further before opening to join formerly separate compartments. However, questions may be raised about this interpretation of what is seen. First, the images produced by thin-section transmission electron microscopy, being artifacts of doubtful relation to the natural

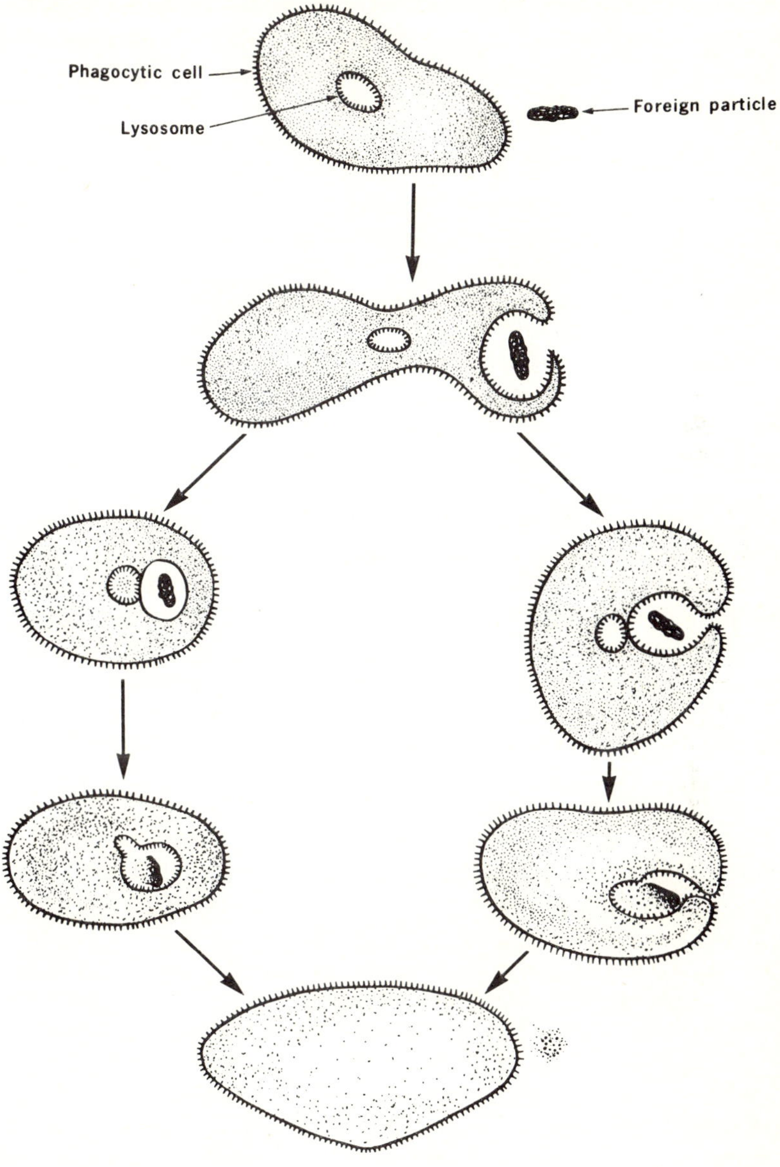

Figure 6-3 Spatial relations of organelles involved in phagocytosis. The left limb of the loop illustrates the case in which fission occurs to form a phagosome. The right limb illustrates fusion with a lysosome without fission of the plasma membrane.

object, vary with the agents and methods of fixation and staining; the number of layers and the spacings vary with the technique. Second, the critical steps in fusion are probably very rapid with respect to fixation, so that distinguishable phases may be missed. Third, there is the unavoidable arbitrariness in assigning temporal relations to static images of a continuous process. Scanning electron microscopy of secretory cells shows pits in the cell surface on the outer face and projections inward from the inner face of the cell membrane, presumably representing the points at which exocytotic fusion has occurred.

Finally in this discussion of exocytosis, it should be recalled (Chapter 5) that the fusion of a secretory vesicle with plasma membrane is a late event in a complex process involving several organelles and cytoplasmic elements. Also, some of the earlier events probably include the fission of membranes. It is a matter of semantic choice whether one uses the term exocytosis to represent the whole sequence or the final event. A similar semantic choice arises in the following discussion of endocytosis.

Two endocytic processes that are long-standing subjects of study in biology are phagocytosis and pinocytosis. Phagocytosis begins with a cellular response to an external stimulus associated with a particle adjacent to the phagocytic cell. The particle may be a bacterium, a bit of carbon, a cell fragment, an aged or abnormal erythrocyte, or something else. Very large phagocytic cells may appear to extend and fold plasma membrane around the particle until the extensions touch and fuse. Others appear to invaginate the plasma membrane where the particle is located outside the cell; the invagination proceeds until the particle is contained in a sac of plasma membrane that fuses at its stalk. These two descriptions are probably of the same process. In either case the sac that is formed, which is a closed piece of formerly plasma membrane, now is an intracellular vesicle called a *phagosome.* (See Figure 6-3.)

(In the preceding paragraph the term fusion is used in a somewhat different sense. The phagocytic process at this stage is a fission in the sense that one membrane has become two. However, fusion has occurred in the sense that the "edges," where membrane has been removed from plasma membrane, reformed a continuous biomembrane.)

The stimuli for this cellular ingestion have been studied with only modest success. Macrophages, derived from mammalian bone marrow, phagocytose many things natural and artificial. Other phagocytic cells, such as neutrophils (another marrow-derived cell), seem more discriminating. It is known that phagocytosis of bacteria is faster when the bacteria are coated with antibody and that the phagocytic macrophages and neutrophils have receptors for antibody in their plasma membranes. Phagocytosis of erythrocytes and some bacteria is sensitive to the carbohydrate on the surface of the potentially ingested cell. Some carbohydrate-coated bacteria resist phagocytosis until the carbohydrate is removed, and intact red cells that are not phagocytosed become susceptible to phagocytosis when sialic acid is removed from the membrane glycoprotein.

Whatever the stimulus, phagocytosis requires Ca^{2+} in the medium surrounding the cell (at least in experimental conditions) and ATP or a competent metabolic energy source. A role for microfilaments is suggested by the observation that cytochalasin B inhibits phagocytosis.

The creation of the phagosome—a vesicle derived from plasma membrane with its glycocalyx facing inward toward the contained foreign particle—is a result of membrane fission. The disposal of the trapped foreign body is a result of membrane fusion. The phagosome membrane fuses with a lysosome membrane. When this occurs, the hydrolytic enzymes of the lysosome and the foreign particle are contained in the same compartment. The several dozen hydrolases digest the ingested particle if it is digestible, like a bacterium or erythrocyte. The breakdown products escape into the cytoplasm. The vesicle formed of the fused phagosome and lysosome (*digestive vacuole*) is usually disposed of, probably by invagination and fission of its membrane to form a *multivesicular body* and subsequent digestion. This sort of fusion may also occur between phagosome and peroxisome, thus exposing the ingested particle to oxidative enzymes. Some morphological evidence has been interpreted as showing fusion of lysosomes with phagosomes-to-be, while they are still continuous with plasma membrane. If this occurs, the breakdown products of lysosomal (and possibly peroxisomal) action may be extruded to the extracellular space. (This has interesting similarities to fusion of a zymogen granule to another that has fused to plasma membrane.)

Pinocytosis is similar to phagocytosis. The ingested material is, rather than a particle, a bit of the external medium presumably containing some of the dissolved or formerly dissolved stimulus of the endocytic process. Again Ca^{2+} and ATP are required, and cytochalasin B inhibits the process. The resulting membrane-bounded intracytoplasmic body is called a *pinosome.* It is usually some tens or perhaps 100 nm in diameter, whereas the phagosome may be micrometers in diameter, depending at least in part on the size of the particle encapsulated. Pinocytosis, in contrast to phagocytosis, seems to be a transport system in some cells, perhaps in most animal cells. For example, the flattened protions of endothelial cells that form the walls of capillaries in mammals are seen to have many invaginations and vacuoles in what appear to be various stages of formation and movement through the narrow dimension of the flattened cell. The various stages of fission or fusion are seen at both luminal and extravascular sides of the cells. This has been interpreted as a mechanism of transport across the capillary wall.

However, the caution about interpreting static images is very pertinent in this case.

Some detailed studies of the uptake of specific proteins by particular cells have shed light on the mechanism of pinocytosis. Cobalamin (vitamin B_{12}) is absorbed from the intestinal lumen after forming a complex with a glycoprotein (intrinsic factor). Intestinal mucosal cell membranes possess receptors for the complex. Whether this uptake is pinocytotic is not known, but the uptake of cobalamin by liver cells and renal cells does seem to be. In the blood, cobalamin is complexed with two proteins (transcobalamins). Hepatic parenchymal and renal cells exposed to a labeled cobalamin-transcobalamin complex can be fractionated. The label is found first associated with a fraction consisting of vesicles apparently derived from plasma membrane and later with a lysosomal fraction. This is quite consistent with pinocytosis.

A case for which more detail is available is the uptake of low-density lipoprotein (LDL) by human fibroblasts in tissue culture. (LDL is a complex of an apoprotein, cholesterol, cholesterol esters, and phospholipids. It is one of a group of plasma lipoproteins that play a major transport role in lipid metabolism.) Fibroblasts have surface receptors for LDL. Apparently these receptors are concentrated in spots on the cell surface, although some are dispersed on the rest of the surface. This distribution can be seen with label made by conjugating LDL with ferritin, a complex of Fe and protein that is identifiable by electron microscopy. These concentrations of LDL are associated with what appears by ultramicroscopic examination to be thickened, depressed (from the outside) regions of plasma membrane. In some cells receptors aggregate after binding of the respective ligand, but in this case binding at a temperature low enough that movement of receptor in the membrane is unlikely still reveals aggregation. Thus, this system may represent a more or less special case in which receptors are aggregated in the absence of ligand.

Since binding of LDL by fibroblasts occurs at low temperature but internalization does not, it is possible to observe internalization by loading the surface receptors with labeled LDL at 4°C and quickly raising the temperature to 37°C. In such an experiment 98% of the label associated with coated regions appears in pinocytotic vesicles (pinosomes) that appear to be formed by invagination of the coated regions of the plasma membrane and subsequent fission. This occurs within 10 min and is followed by movement of the vesicles within the cytoplasm, a disappearance of the so-called coat at the cytoplasmic face of the vesicle membranes, and fusion with lysosomes.

The proportion of membrane in the coated regions is small, around 1-2% of the cell membrane. Clustering of LDL-receptor complexes in these regions and preferential pinocytosis at these points probably explain the rate of internalization of LDL-ferritin. In an hour the fibroblasts internalize twice as much as that bound to membrane at any one time. So, if the uptake were random over the surface and the distribution of receptor were random, the plasma membrane would have to turn over twice in 1 h to explain the rate of internalization. In other investigations membrane turnover in these cells has been estimated at one-fourth of the cell membrane in 1 h. If pinocytosis were the only process of internalization, then, it would be inadequate by a factor of eight to explain the rate of uptake with saturated receptors—under the assumption of random pinocytosis. Therefore the measured rate almost certainly results from aggregation of LDL-receptor complexes and location-specific pinocytosis at the points of concentration of complexes.

One more process must be considered in the interpretation of these studies. Both liver cells with respect to transcobalamin-cobalamin complex and fibroblasts with respect to LDL show a low rate of pinocytotic uptake that is nonspecific in the sense that identifiable surface receptors are not found to be involved. This process is orders of magnitude slower than the receptor-mediated pinocytosis and is not saturable at typical concentrations of the exogenous substrate. Nevertheless this apparently nonspecific binding probably (as suggested by label studies) contributes to the major process through migration of membrane-bound substrate to the specific sites of pinocytosis.

6-5 MITOSIS

The division of a eucaryotic cell into two genetically complete daughter cells is most often studied with respect to the replication and segregation of chromosomes (*karyokinesis*). However, two membrane systems, the nuclear envelope and the plasma membrane, play important parts in this very complex, integrated process. The nuclear envelope's role is to divide or to get out of the way, depending on the particular cell. The plasma membrane, on the other hand, has two roles: it initiates, or permits initiation of, mitosis, and it participates in that part of mitosis involving the division of cytoplasm and cell membrane (*cytokinesis*).

The behavior of the nuclear envelope during mitosis varies among eucaryotic cells. In some cells the nuclear envelope divides as a recognizable structure. In mammalian cells undergoing mitosis, the nuclear envelope disorganizes to the extent that the nuclear membranes appear as fragments during prophase. The pores of the nuclear membrane are not found in late prophase. These fragments may merge with endoplasmic reticulum. During telophase the nuclear envelopes of the daughter cells form while the chromosomes are still in the condensed form characteristic of mitosis. At this time the cisternae of endoplasmic reticulum are aligned around the chromosomes, and the forming nuclear envelopes are probably derived from endoplasmic reticulum and fragments of the nuclear envelope of the parent cell. The fusion of these membranes and the reappearance of nuclear envelope pores mark the completion of karyokinesis.

The plasma membrane undergoes a complex series of changes during cytokinesis. In most mammalian cells these changes are synchronized with the stages of karyokinesis. Cytokinesis begins with the formation of a division furrow that is located where the plane of symmetry of the spindle apparatus intersects the plasma membrane. The spindle apparatus determines the location of the division furrow (in cells such as sea urchin oocytes where the manipulations can be done), as indicated by its moving when the mitotic spindle is moved. However, once initiated, cytokinesis progresses in the absence of the spindle. The relation between the spindle and the cell membrane may be mediated by the microtubules of the asters and spindle fibers that radiate outward toward, sometimes coming into contact with, the cell membrane.

The division furrow is a circumferential indentation around the cell. It is associated with a so-called contractile ring within the cytoplasm adjacent to the furrow. The contractile ring is composed of microfilaments (identified by electron microscopy) and myosin (identified by fluorescent antibody). The formation and development of the division furrow require the presence of Ca^{2+} and can be maintained in the absence of normal cytoplasm by Ca^{2+}, Mg^{2+}, and ATP. The contractile ring appears with the furrow as bundles of microfilaments parallel to the membrane. The indentation (furrow) deepens, and the contractile ring decreases in circumference (apparently without a change in cross-sectional area) and eventually disappears as the cell divides into two daughter cells. The membrane of the furrow remains closely associated with the contractile ring, forming folds perpendicular to the furrow—and microvilli in some cases—as the furrow deepens and the waist of the cell (or joined cells) narrows. The appearance is that of a purse string pinching one sac into two, shortening without thickening as if dissipating as it tightens.

Any change in surface-to-volume ratio that accompanies cytokinesis is accommodated by shape change. This sort of accommodation is suggested by the greater density of microtubules and folds on cells before mitosis than on smaller daughter cells immediately after mitosis.

Biochemical changes in the plasma membrane accompany mitosis. Or, perhaps they are not so much biochemical changes in the usual sense as they are changes in the exposure or mobility of unchanged parts of membrane components. A particularly interesting example is the agglutinability of cells at different stages of the cell cycle. Plant lectins (proteins with specific affinities for cell surface receptors) react, in most known cases, with particular sugar or substituted sugar groups of membrane glycoproteins. If the lectin used is multivalent it induces agglutination of the cells to a much greater degree during mitosis than during interphase. This indicates an altered orientation or conformation of glycoprotein, an increase in receptor number, or an increased mobility of glycoproteins. (Translational mobility of glycoprotein receptors in the membrane is prerequisite to lectin-induced agglutination.)

A possibly related biochemical change is seen in the enzymatic activity of certain plasma membrane-bound glycoslytransferases characteristic of the plasma membrane. These enzymes are active with respect to the medium outside the cell. In tissue culture some cells exhibit a variation in surface glycosyltransferase activity in phase with the cell cycle, so that this activity is maximum during mitosis and minimum during interphase. Comparison of cultured cells in suspension with those fixed to a surface suggests that interphase cells glycosylate adjacent cells, whereas during mitosis the glycosylation is of membrane glycoproteins by the enzyme of the same cell membrane.

The increase in lectin-induced agglutinability during mitosis must be related to the mitogenic properties of some lectins for some cells, but the connection is not yet understood. These lectins induce mitosis in some cells, including mammalian cells. Other agents induce mitosis as well. This mitogenesis by exogenous, impermeable agents makes it quite clear that the plasma membrane—in addition to its role in cytokinesis—controls the timing of mitosis.

The case for the plasma membrane as mediator between exogenous mitogen and cell division is not without weak points. For example, both lectins and proteases (such as trypsin, chymotrypsin, and pronase) that induce transformations leading to mitosis in lymphocytes and fibroblasts require a day or days of exposure before mitosis is induced. Under such circumstances endocytic incorporation of the inducer-receptor complex is not out of the question. However, lectin fixed to a solid surface still induces mitosis. The effect of the inorganic mitogen sodium periodate (which requires exposure of only 10 min) can be reduced or eliminated by reducing agents or by pretreatment of lymphocytes with neuraminidase (an enzyme that removes the sialic acid group from the carbohydrate portion of glycoprotein). These observations seem to show that specific surface receptors mediate mitogenesis and that incorporation of the mitogen into the cell interior is not required.

Both lectins with lymphocytes and fibroblasts in culture, and periodate with lymphocytes in culture, require serum factors to initiate mitosis. Proteases induce mitosis without serum factors. The possibility that the long exposure required is associated with internalization of the enzyme is not as surely refuted as in the case of lectins, but it is worth noting that proteases enhance the mitogenic effect of lectins. This is consistent with a membrane-mediated response dependent on integral membrane protein interactions.

Under natural conditions cells surely move through the cell cycle, and into mitosis, in response to immediate environmental conditions. There are clear examples in simpler species. As a case in point, yeast cells bud when one provides water and nutrients in making bread. Cells of metazoans, however, must respond to a more organized environment. The requirements for Ca^{2+} and for the exogenous precursors of ATP exist, but even in the presence of basic requirements cells of metazoans do not divide uncontrolled. There are both stimulants to, and inhibitors of, mitosis.

Several natural mitogens have been identified in mammalian systems. These factors are the components of serum required in the artificial situations mentioned above. Nondividing cells in culture can be induced to divide with these substances, called *growth factors*. Some of them—perhaps any of them sufficiently pure—are specific for the target cell. Nerve growth factor and epidermal growth factor are prominent examples. These growth factors are peptides composed of some tens or 100 amino acids. They act by binding to receptors on the cell surface; the evidence suggests that their entry into the cell is not required for this effect, although proof is difficult to obtain when such small amounts can be effective. (See the discussion of hormone action in Section 7-4.) Whatever the mechanism, the presence of receptors on the plasma membranes of metazoan cells makes it possible for cells to respond to signals from distant cells and other tissues by undergoing mitosis.

To complete the general picture, inhibitors of mitosis must be mentioned. Some of these inhibitors are hormone components of a feedback system involving a pair of secreting glands (for example, the anterior pituitary-adrenal gland system). More recently, substances that are inhibitors of mitosis in the same cell as that which produces them have been discovered. Called *chalones,* these substances reversibly inhibit mitosis—both in cultured cells and in cells *in situ.* It may be quite generally true that the plasma membranes of animal cells capable of mitosis are continuously monitoring a field of mixed stimulants and inhibitors that compete to induce or retard movement through the cell cycle.

6-6 SUMMARY

The number of separate biomembranes in a cell or group of cells can change by fusion or fission. Fusion of two membranes reduces the number of separate compartments as well as the number of biomembranes. Fission of one biomembrane into two increases the number of compartments.

The end states of a fission or a fusion event are states of minimum free energy relative to the transition state. Fusion includes the following steps: contact, transition, and shape change; fission includes shape change, transition, and separation.

All fission and fusion events with biomembranes preserve the orientation of membrane components so that the glycocalyx (and usually the amino-terminal end of the peptide chains of membrane-spanning integral proteins) is on the side of the membrane opposite the cytoplasm.

Exocytosis is a process in which intracellular vesicles fuse with plasma membrane, exuding their contents into the extracellular space. Examples of this are secretion of enzymes, salts, and neurotransmitters, and biogenesis of plasma membrane. Endocytosis is a process by which extracellular material is brought into a cell by invagination or extension of plasma membrane that undergoes fission to form an endocytic vesicle. Examples are recycling of plasma membrane, pinocytosis, and phagocytosis.

Mitosis involves coordinated membrane changes. The nuclear envelope fragments and reforms into two envelopes in mammalian cells. The plasma membrane responds to karyokinesis by going through a complex series of changes in cytokinesis, which ends with fission into two separate plasma membranes.

Chapter 7

Biomembranes and Intercellular Relations

The plasma membrane is the interface of cell with environment. For any given cell the environment includes–in metazoans–both extracellular fluid and other cells. In some cases the environment also includes basement membrane or other wet, solid extracellular matrix. The cells of the environment may be subdivided into those in contact with the cell under consideration and those at a distance. Cells in contact may communicate by transfer of chemicals, pressure, electrostatic force, and electrodynamic forces. Distant cells may communicate by transfer of chemicals. Of course, two widely separated cells may communicate through other cells interposed between them, but this process is simply a summation of the other two.

Preceding chapters have necessarily included some aspects of chemical communication between cells. Section 7-1 is a discussion directed more specifically toward the intercellular transfer of information and cellular responses to chemical messengers. Section 7-2 is a brief presentation of some general aspects of agglutination of cells, which is a relatively primitive kind of cell-to-cell contact. In Section 7-3 the roles of cell contact and cell response to chemical gradients in the development of multicellular systems are presented. Section 7-4 is a description of specific types of cell-to-cell junctions and their roles in multicellular systems.

7-1 CHEMICAL COMMUNICATION BETWEEN CELLS

In Section 3-2 neurotransmitters were mentioned in connection with excitable cells, and in Section 6-5 growth factors and chalones were mentioned in connection with the initiation of mitosis. These are examples of the chemical transfer of information describing the status of the extracellular medium and/or the status of adjacent or distant cells. The history and context in which each class of chemical messenger was first discovered and defined have led to some artificial distinctions. The emphasis here is on the common features.

Common characteristics of the chemical messengers to be discussed in this section are that they are natural, specific, and endogenous with respect to the organism. This excludes (on the grounds of specificity, at least) nutrients that do in fact carry information to a cell and exercise control over the status and activity of the cell. Glucose, for example, is not considered here to be a special chemical messenger that, via ATP, controls the activity of the Na^{+}-K^{+} exchange transport of the erythrocyte membrane.

The remainder of this section is a discussion of hormone action. Hormones play a prominent part in organ physiology. They form the link between activator (secretor) tissues and activated (target) tissues. The effects of hormones on organs and organisms result from alterations in the activities and products of target cells. Whatever these cellular responses may be, they follow either an interaction of the hormone with a receptor in the plasma membrane or the hormone's crossing the plasma membrane and entering the cell.

Many mammalian hormones are now known. They may be classified according to chemical nature into two groups: (1) sterols or small molecules and (2) peptides. It has been thought that these two chemical groups correspond conveniently to two kinds of interactions with the target cell. The sterols appear to cross the plasma membrane without a specific receptor reaction, bind to a cytoplasmic receptor molecule, enter the nucleus, and thereby alter the cell's activity. The peptide hormones, on the other hand, were thought to react with plasma membrane receptors (never crossing the membrane). This reaction would induce the release of a so-called second messenger, most commonly by activation of adenyl cyclase on the cytoplasmic side of the membrane, thus releasing cyclic adenosine monophosphate (cAMP), which triggers a variety of changes in intracellular enzyme activities. This two-class conceptualization of hormone action is a useful starting point, but it is an oversimplification.

There are small molecules that are not sterols but that do function like hormones. These include the various neurotransmitters such as acetylcholine and vasoactive substances such as serotonin (5-hydroxytryptamine). These are often not considered among the hormones, being less specific with respect to secreting cells and/or target cell response. They probably all exert their influence by interaction with the plasma membrane—as the neurotransmitters do—which makes them more akin to the peptide hormones with respect to target organelle.

The sterols are relatively quite lipid-soluble and probably normally cross the plasma membrane nonspecifically. The sequence of intracellular events leading to the cellular response involves cytoplasmic reactants (soluble receptors) and specific effects on the genome, presumably derepression or repression of genetic loci associated with specific protein synthesis. The significance of the biomembrane in such a hypothesis is the regulation by the nuclear envelope of the relation between cytoplasmic hormone-receptor complex and the ultimate target within the nucleus. In contrast, the role of the plasma membrane in regulating peptide hormone action includes a variety of mechanisms rooted in the very nature of a biomembrane.

Receptors for specific peptide hormones are found associated with plasma membranes of target cells. A receptor for triiodothyronine, a product of the thyroid gland that increases oxidative phosphorylation, is found associated with the inner mitochondrial membrane. (Triiodothyronine is an amino acid; it is small and has two benzene rings. It shares properties of the two classes of hormone and apparently reacts specifically with a membrane but in this case with an intracellular membrane.) Insulin, the hormone lacking in the disease diabetes mellitus, has target cells of various types. It has been studied extensively, and what has been learned illustrates well the role of the membrane in determining cellular response.

Human insulin is composed of two polypeptide chains joined by disulfide bonds into a complex of 51 amino acids (derived from a larger polypeptide called proinsulin). It binds to receptors in the plasma membrane of target cells (and to cells not usually considered targets for insulin). The specific receptor is a glycoprotein, an integral membrane protein. The binding of insulin is the first of a series of distinct events that produce the responses of the cell. Insulin may be covalently bonded to large molecules to form complexes that are unlikely to cross the plasma membrane but that retain physiological activity. There is some question at to the permanence of these complexes; it is possible, because of the very small amounts of hormone required for a physiological response, that the activity is a consequence of the breakdown to free insulin. The evidence in the case of another peptide hormone, nerve growth factor, is better. In this case the site of binding of the hormone-label complex corresponds to the site of hormone action at points of localized nerve cell growth. Such observations lead to the idea of a transducer in the membrane, either the receptor glycoprotein or an associated membrane component, which conveys the signal from the binding site at the extracellular face of the membrane to the cytoplasmic face of the membrane.

The plasma membrane thus has three roles to play: it carries the receptor and influences both the number of receptors and their affinity for the hormone; it carries the message "hormone bound" to the cytoplasm or cytoplasmic face of the membrane; and it is the effector for some cellular responses to specific substrates, such as altered cell membrane permeability.

The first step in the cellular response is the binding of hormone to receptor in the plasma membrane. The extent of binding, for a given extracellular hormone concentration, depends on two factors: the number of receptors and the affinity of the receptor for the hormone. Observations of insulin binding in several mammalian cell types indicate a variation with extracellular hormone in the number of receptors per cell accessible for binding. In the presence of a high concentration of hormone, the number of receptors decreases (down-regulation) over minutes, and when the extracellular medium is replaced with hormone-free medium the number of receptors increases over hours. Microscopic observation of insulin bound to fluorescent labels shows a clustering and internalization of label—presumably of the insulin-label-receptor complex. An explanation that fits nicely with the ideas of membrane synthesis and turnover (Chapter 4) and with receptor-mediated pinocytosis (Section 6-4) is that the binding of hormone to receptor is followed by the invagination of areas of plasma membrane that include aggregates of hormone-receptor complex. This, in turn, is followed by endocytosis, fusion of endocytic vesicles with lysosomes, degradation, and finally recycling of internalized material. Such a process would provide a stabilizing negative feedback for cellular response to hormone and—coupled with synthesis of hormone receptor—would keep the cell in a state of responsiveness to changes in hormone concentration. Such a continuous synthesis of receptors is suggested by the observation that proteolytic destruction of insulin receptors is followed by their reappearance over several hours.

The matter of variation in affinity of receptor for hormone (variation in the value of the equilibrium constant for the binding reaction) is an open question. Evidence has been presented for a decrease in affinity with increasing hormone concentration, but this evidence comes from difficult experiments and perhaps tenuous results, and the explanation of the data may not require the assumption of a changing affinity. There is precedent for allosteric effects in complexes of proteins that change binding affinities (hemoglobin, for example). There are also known cooperative effects in ligand-binding proteins. In the case of insulin binding by erythrocytes, there is evidence of a transition of receptors from tetrameric form to monomeric form with an apparent decrease in affinity for insulin.

The second of the three roles of the plasma membrane is that of conveyor of information from outside the cell to the cytoplasmic side of the membrane. Less is known about this step than about the binding of hormone to receptor. For a number of peptide hormones the association between an intracellular increase in cAMP and the binding of the hormone to plasma membrane has led to the hypothesis that a transducer protein in the membrane activates membrane-bound adenyl cyclase, which catalyzes the reaction ATP → cAMP + PP (pyrophosphate). The cAMP released in the cytoplasm then serves as a second messenger in altering the activity of several intracellular enzymes that mediate the response to the hormone. These particular intermediates in the sequence probably do not apply in the insulin case; decrease, or no variation, in cAMP has been reported in response to insulin. Nevertheless the circumstantial evidence suggests a second messenger. Ca^{2+} has been suggested as a general second messenger.

Adenyl cyclase itself is a lipoprotein that depends on a lipid environment to maintain its enzymatic activity. In fact, reagents that alter membrane lipids reduce the adenyl cyclase activity of membranes containing the enzyme. Extracted adenyl cyclase is restored to activity by specific phosphoglycerides in some cases.

The third role of the plasma membrane is that of effector of the cellular response. Insulin binding, for example, is followed by increased transport of glucose and amino acids into target cells. The mechanism of this response is not known. Aggregation of label on cells exposed to labeled hormone has been observed, leading to the suggestion that hormone-induced permeability changes are a consequence of redistribution of receptors and perhaps other membrane proteins.

The intracellular responses to peptide hormones are generally seen as biochemical changes. Insulin increases protein synthesis and decreases lipid degradation in target cells. In addition insulin modifies responses to changes in concentrations of other hormones. Such biochemical effects may be manifestations of either cytoplasmic interactions or intranuclear interactions with the genome. In the context of the picture of hormones as fitting into two groups, one acting on plasma membrane and the other on the nucleus, it is tempting to concentrate on cytoplasmic effects to explain responses to peptide hormones. However, it must be kept in mind that the complex and dynamic state of the cell includes interaction among cytoplasmic components, extranuclear organelles, and nuclear contents.

The picture is complicated somewhat by findings of intracellular receptors for peptide hormones. There are reports of insulin receptors on lysosomes, endoplasmic reticulum, and nuclear envelope. Again the experiments are difficult, and some attempts to repeat them have not shown the receptors where they had been seen before. Perhaps the most important question raised by such observations is that of just where insulin, for example, works. Are any of the intracellular responses caused by intracellular binding of hormone? The question is unresolved at the moment, but the mechanisms proposed in Chapter 4 for biogenesis and turnover of membranes—for which there is good evidence—would explain the presence of receptors on membranes of intracellular organelles. The glycoprotein that is the insulin receptor must be synthesized somewhere, and that place is not likely to be the plasma membrane.

7-2 CELLULAR AGGLUTINATION

The distinction between chemically mediated cellular interactions discussed in Section 7-1 and the contact interactions discussed in this and the following sections should not be overemphasized. Cells in contact exchange chemical messengers, of course; but the contact itself, if it leads to a stable association, is a chemical event. The distinction between a peptide hormonal message between cells and fixation of one cell to another is that a dissolved chemical reacts with a membrane-bound receptor in the former case, whereas two membrane-bound "receptors" react with one another or with the same molecule in the extracellular environment in the latter case.

The terminology of cellular aggregation is not entirely uniform. *Aggregation* can mean the formation of clumps of cells of any kind. It, rather than *agglutination,* is also applied to highly organized and specific groups of cells. Agglutination, on the other hand, is used more often to describe nonspecific clumping or the aggregation associated with immune responses and blood clotting. In this section agglutination is used in this more or less general sense; the aggregates are disorganized at the cellular level.

With agglutination, as with many biologic phenomena, the simpler cases can be produced in the laboratory where the process is artificial but to some extent controlled. Cells that are normally free in extracellular fluid do not agglutinate *in vitro*—or do so extremely slowly—if kept in a physiological medium. They will under some circumstances become fixed to an artificial substrate such as glass or plastic. With both cultured cells and mammalian erythrocytes this fixation depends on the presence of a serum factor or secreted substance or a suitable substrate. This factor can be a protein that forms a coat on the substrate.

The fixation of cell to cell by artificial means has been observed carefully with human red cells. One of the more interesting investigations was a study of the effect of polymer chain length on agglutination. Polylysine can be separated according to chain length [that is, number of monomer units (lysine) per polymer molecule]. Agglutination occurs only when the chain length is great enough, and the degree of agglutination is greater for greater chain length at a given concentration. Electron microscopy reveals cell separations corresponding to the extended chain lengths of the polymer molecules. The acidic polymers do not cause agglutination; the basic ones do.

Erythrocytes demonstrate a special form of aggregation that can be induced with both natural and artificial agents. They form rouleaux, which are groups of cells that resemble a stack of coins. The separation between the cells of a rouleau is approximately 20 nm, as in other nonspecific aggregates. Rouleau formation occurs naturally in sludged blood and *in vitro* is apparently the major determinant of the erythrocyte sedimentation rate (ESR), which is a common diagnostic test. Rouleaux do not form from a suspension of erythrocytes in a protein-free physiological salt solution. Dextrans (branched polysaccharides of glucose) and the polymer polyvinylpyrrolidone (PVP) induce rouleau formation. Among the natural inducers is fibrinogen, the plasma precursor to fibrin that appears during clotting. If the plasma (with added anticoagulant) of blood that exhibits rapid rouleau formation is mixed with red cells from blood that does not, rapid rouleau formation occurs. If the red cells from the rouleau-active blood are mixed with plasma from the rouleau-inactive blood, rouleau formation is not remarkable. The inducer is in the plasma, not on the cells.

Another fascinating observation was made with erythrocytes, in a case of agglutination induced by the plant lectin concanavalin A. Concanavalin A will agglutinate human red cells when it is in a multivalent form and at modest concentrations under conditions in which the integral membrane glycoproteins are mobile in the membrane. When concanavalin A is fixed to the surfaces of plastic microspheres, red cells stick to the surfaces of the spheres even when there is no detectable free concanavalin A in solution. The interesting observation here is that there is not only one layer of red cells covering each microsphere; there are several layers of cells—cells not in contact with the microsphere but nevertheless stuck. The lectin-induced agglutination seems to pass from cell to cell.

Most of the investigation of contact and fixation of mammalian cells has been done with embryonic tissues. Some of the results are discussed in Section 7-3. There are, however, some pertinent physiological (as opposed to embryologic) phenomena that have been observed in considerable detail. One of these is the clotting of blood.

Blood clotting is a complex series of events involving soluble plasma components that undergo dozens of reactions with feedback modulations and elaborate responses of thrombocytes (blood platelets). The role of the biomembrane is of interest here. Platelets have no nucleus but do contain other membranous organelles and have a typical plasma membrane. Platelets contribute to clotting in several ways, but the early response that is particularly related to the plasma membrane is a precipitous appearance of adhesiveness. Apparently disruption of the blood vessels causes adhesion of platelets to tissue in the region of the wound and to one another. This reaction has been attributed specifically to contact of platelet with collagen, but later work seems to indicate that some other factor is needed.

Clumping of platelets seems to require a certain minimum mechanical agitation; it does not occur in undisturbed blood in the laboratory. Agents that induce clumping in the presence of Ca^{2+} and probably a plasma protein (perhaps fibrinogen), which are required, include collagen, saturated fatty acids, ADP, norepinephrine, 5-hydroxytryptamine, and thrombin. Removal of Ca^{2+} after agglutination leads to separation of the cells. Agglutination induced by minimal amounts of ADP will reverse spontaneously, and AMP blocks induction by ADP. The aggregation response *in vitro* occurs in two phases in response to some agents. The first clumping begins in a second or two and is complete in a half minute or so. At about 1 min after adding the inducer, the second phase begins and is complete in another minute. The second phase is associated with the so-called release reaction in which platelets release ADP, serotonin, and other substances, at least some of which contribute to blood coagulation. Collagen causes a delayed one-phase agglutination.

Detailed studies of lymphocytes have also produced much information. The immune response includes interactions among two types of lymphocytes and sometimes macrophages that progress through various stages. The interactions involve surface components in the plasma membranes. Lymphocyte plasma membranes include immunoglobulins that act as receptors for antigens. The antigen-receptor reaction stimulates mitosis and antibody synthesis. In addition, the immunoglobulins of the cell surface are responsible for interactions between cells. One may not want to regard this as an example of agglutination, but it is a useful case for clarifying the point that at least some intercellular contact interactions are very specific. The lymphocyte interactions are mediated by immunoglobulins that are as specific in their binding properties as enzymes and more so than any other class of receptor or ligand.

An interesting feature of lymphocyte behavior that seems to be generally relevant to intercellular agglutination is *capping* (Figure 7-1). This is a phenomenon that accompanies antigen-induced activation of lymphocytes. Antibodies against the surface immunoglobulins of lymphocytes can be conjugated with microscopically detectable labels while remaining effective lymphocyte activators. The binding of these labeled (divalent) antibodies

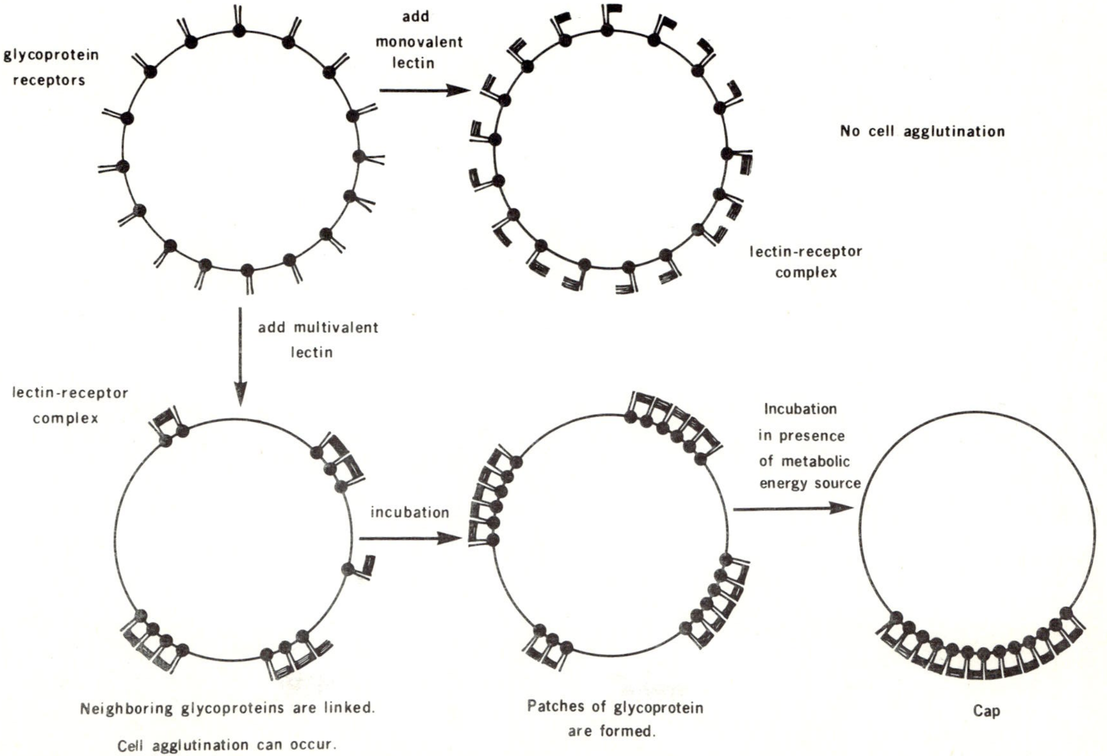

Figure 7-1 Lectin-induced clustering and capping of cell surface receptors. With some cells the cap is internalized by endocytosis.

to the immunoglobulin in the plasma membrane is followed by the formation of patches of label on the cell surface. This patching can occur in metabolically inactive cells. If the cells are sufficiently warm and metabolically competent, the label further aggregates into a cap on the pole of the cell opposite the nucleus, which is normally off center. This is followed by pinocytosis of the label.

All this is reminiscent of the discussion of hormone action and the possible role of receptor redistribution (Section 7-1) and receptor-mediated pinocytosis (Section 6-4). The relevance to agglutination is that this patching often—perhaps always—accompanies agglutination with agents that bind to membrane glycoproteins. Maneuvers that prevent patching also suppress agglutination. A mechanism that includes the redistribution of integral proteins may well explain such propagated local membrane changes as may underlie agglutination of free cells to agglutinated cells (*vide supra*) and fusion of secretory vesicles to vesicles just fused to plasma membrane (Section 6-4).

7-3 CELL SURFACE INTERACTIONS IN DEVELOPMENT

Biologic development may be examined as a coordinated combination of three processes: cell differentiation, morphogenesis, and growth (although growth may be considered an aspect of morphogenesis). Cell differentiation involves all the complexities of molecular genetics. It is discussed here only in terms of changes in the cell surface. Morphogenesis involves cell-to-cell associations and cellular responses that govern mitosis, cell motility, cell junctions, and the control of these cellular processes in such a way that predetermined sizes, shapes, and compositions of tissues and organs result. This section is not a discussion of development so much as it is an effort to relate the role of the plasma membrane in metazoan development to the basic characteristics of biomembranes presented in preceding sections.

In the mammal the development of blood is a good example of differentiation, because morphogenesis is not involved in the biogenesis of the various types of free, single cells. Undifferentiated stem cells reside in the bone marrow where they divide and differentiate in response to blood-borne hormonal signals. These cells, in turn, divide and differentiate along the lines that end in erythrocytes, leukocytes, and platelets. Because of its importance in medicine, knowledge of the morphological stages of this differentiation has been thoroughly developed. The cells derived from this common origin are strikingly different. The various leukocytes (white blood cells) are motile, active, nucleated cells that respond to their environment in complex ways in the antigen-antibody reaction, by ruffling, and in phagocytosis. The thrombocyte (platelet) has no nucleus but responds in complex ways to environmental changes usually associated with blood clot (thrombus) formation. The erythrocyte, perhaps the simplest of cells, has no nucleus, possesses a minimum of metabolic apparatus, and has a dry mass over 90% of which is a single protein-prophyrin-Fe complex (hemoglobin). The stem cell generates by mitosis and differentiation several cell lines each of which undergoes a series of mitoses and differentiations.

As an example of biomembrane differentiation, consider the erythroid line of differentiating cells from human and rat bone marrow and circulating blood. These cells have been labeled with concanavalin A and colloidal Fe. Concanavalin A was conjugated with an identifiable label for microscopic location. The colloidal Fe was in the form of a detectably large, positively charged clump of oxide. There is a progressive increase in concanavalin A binding and a progressive decrease in colloidal Fe binding with the successive division of erythroblasts (an early erythroid stage). Then, as the orthochromatic erythroblast (a later stage) differentiates further and becomes a reticulocyte (the immediate precursor, having no nucleus, of the mature erythrocyte), there is decreased binding of convanavalin A and increased binding of colloidal Fe. The greatest change occurs when the nucleus is expelled. The nucleus carries with it some plasma membrane that is several times more heavily labeled with concanavalin A and several times less heavily labeled with Fe than the remaining plasma membrane. Thus, as the cells differentiate, either new glycoproteins are introduced into the plasma membrane or new modifications are introduced into an ongoing synthesis of glycoproteins of the plasma membrane. Also, differentiating cells take on new behaviors (functions). In many cases these observations are clearly related—that is, cell behavior is causally linked to the composition of the plasma membrane.

The single cell, presumably, has differentiated to a state in which it has particular functional capacities. Its surface includes receptors and antigens related to this end state. All it must do up to that point—as far as differentiation of its surface is concerned—is to avoid getting stuck in one location. (Stem cells and differentiating blood cells do stay in the marrow until a particular stage is reached and then enter the circulation. Nevertheless, the contrast with cells that are fixed is probably valid.) On the other hand, cells that are, or become, parts of solid tissues and organs undergo surface differentiation that plays a major role in the process of development itself. The earliest signs of this role in mammals are seen in embryogenesis.

One of the similarities of the early mammalian embryo to the simpler (more "primitive") species is the relatively undifferentiated and totipotential nature of the component cells. Another similarity is revealed in observations of cell sorting. Simple metazoans such as sponges have been treated with enzymes, or forced through a mesh, or mechanically disrupted until suspensions of single cells and/or fragments smaller than whole organs remained. In time (hours to days) these cells and fragments reorganize into normally oriented tissues and organs. Vertebrate embryos behave similarly.

In stages late enough that the organs have developed, separated cells reaggregate to a state recognizably like the organs from which they came. Tubules form from cells in culture taken from kidney, for example. The sorting out and reaggregation follow both tissue specificity and species specificity, but tissue specificity is manifest over species specificity. In one case, mouse heart cells sorted out with cells derived from chick heart in the presence of mouse cells from nonheart tissues.

These relatively crude—but seminal—observations have led to an emphasis on what is called *intercellular adhesive specificity*. This concept, as an explanation of cell sorting and species- and tissue-specific aggregation, has led to more refined observations that define the surface components responsible for intercellular adhesive specificity. This specificity is seen to vary with the stage of development in a given embryo—as might be expected from what is known of the branching that occurs during embryogenesis and organogenesis.

Cell sorting, of course, is not a natural process except in organisms such as *Dictyostelium discoideum* (a slime mold) that have life cycles that include free-living single cells that aggregate to form a multicellular, multitissue organism. A very similar slime mold can be mixed with *D. discoideum* in the single-celled stage. Then, when conditions that induce aggregation are imposed, the cells aggregate and separate according to species. The chemical components of slime mold cell membranes responsible for aggregation have been isolated and studied. They include at least one glycoprotein.

Cell migration is a natural part of embryogenesis. The process probably begins in some cases with a change in cell surface that releases the cells from a specific adhesiveness and ends with a specific adhesion to another group of cells in another region of the embryo. In chick embryos single primordial germ cells migrate through the vascular system and end their migration in association with the gonadal mesenchyme. In another avian system lymphoid cells always migrated to a thymus implant no matter where the implant was located. Neural crest cells also go through a migratory phase ending in specific associations. Nerve growth factor (Section 7-1) is required for this adhesive specificity; it also induces differentiation and growth in specific cells.

Migration and specific adhesion occur also in solid embryonic tissues. Neuroblasts migrate through portions of the developing nervous system to form specifically located layers of specialized cell types. Axonal tips also progress with growth through solid tissue. Experimental studies have shown that the elongation and adhesion of the tips depend on the chemical nature of the substrate on which they grow. The intricate control exhibited when innervating pathways form in embryos seems to show that such contact-responsiveness is operative in the embryo.

The exact chemical components involved in these intercellular interactions are not yet known in mammalian systems. However, the class of component is known with qualified certainty in some cases. An increase in identifiable carbohydrate-containing macromolecules is found to precede the joining at the midline when the palate is formed. With a cultured rat muscle cell, formation of myotubes by cell fusion is blocked by thiodigalactoside, which would block galactose binding. These cells contain a carbohydrate-binding protein when capable of fusion. In many other studies of specific adhesion in embryogenesis, evidence for the role of carbohydrate binding and of glycoproteins has been found. With what is known about glycoprotein synthesis, biogenesis of plasma membrane, and epigenetics, it is inviting to imagine a temporally unfolding embryogenesis that is mediated by sequential changes in the plasma membrane and programmed in the genome.

More direct evidence exists for the role of glycoproteins in adhesion. Chick cells dissociated by treatment with trypsin are kept from reaggregating by exposure to periodate, which degrades the carbohydrate groups of glycoproteins. Glucosamine, which will block some reactions with surface carbohydrate groups, causes chick cells to reaggregate from larger into smaller clumps. Hexosamine synthesis is found to increase in parallel with intercellular adhesion in a suspension of mouse cells. A lack of glutamine and competition of a glutamine analog are associated with diminished aggregation; the effects of both are reversed by hexosamine. In retinal nerve cells blockage of hexosamine synthesis is associated with both reduced glycoprotein synthesis and reduced reaggregation.

Efforts to isolate and identify factors responsible for aggregation in embryonic tissues have proved interesting. A glycoprotein from the medium in which chick retinal neurons are growing specifically stimulates adhesion of these cells. A plasma membrane protein from these cells inhibits aggregation.

Some suggestive observations have been made with rat liver cells. As embryonic age increases the glycosyltransferase activity that adds mannose to glycoprotein decreases. In contrast the activity for addition of galactose is 5 times greater in 17-day-old embryos than in 35-day-old rats. Adhesiveness changes as well. In embryonic retina the sorting and aggregation described above have been shown to be accompanied by the appearance of glycoprotein on cell surfaces.

In addition to a temporal pattern in glycosyltransferase activity, there is a spatial pattern. In an early stage of embryogenesis–the gastrula–there seem to be different specific glycosyltransferase activities in different regions.

These relations between adhesiveness and glycosyltransferase activities have been interpreted to mean that the membrane-bound glycosyltransferases serve a receptor or binding function for markers on other cells as well as an enzymatic function. Binding is, after all, a first step in enzymatic activity. It may be that these enzymes stop at the binding step under certain conditions. The critical conditions may have to do with the chain length of the bound substrate, which can be a carbohydrate of a glycoprotein of an adjacent cell. It is not yet clear that this mechanism for specific, controlled intercellular adhesion is in fact a real one. In any case the idea is a useful working hypothesis.

7-4 INTERCELLULAR JUNCTIONS

In the introduction to this chapter it was pointed out that cells in contact may interact by transfer of chemicals and by mechanical, electrostatic, and electrodynamic forces. These possibilities are common to physical systems. Living cells, however, exhibit very special contact interactions that have peculiar morphological and functional characteristics. The discussion of cellular interactions in development in the last section makes it quite clear that the organization of cells into tissues and organs is quite specific and coordinated. So, in considering *intercellular junctions* (or *junctional complexes*), one may proceed from the fact that cells are fixed in relation to one another in very specific and orderly ways.

The functional outcomes of such fixation are (1) orientation and fixation of cells, (2) formation of extracellular compartments, and (3) creation of special connections between cytoplasms of adjacent cells. Each of these functional characteristics of cell-to-cell fixation has a structural correlate in the form of a localized specialization of plasma membrane.

The structural correlate of simple fixation of cell to cell is the *desmosome,* also called the *macula adherens* (see Figure 7-2). The adjacent plasma membranes of the two cells maintain their usual cross-sectional appearance and the separation of about 20 nm, which is characteristic of mammalian cells in close association. In the desmosome region, however, the intercellular space is relatively opaque to electrons after the usual OsO_4 staining. On the cytoplasmic side of either membrane there is also a dense region about 20 nm thick that occupies an area coextensive with the intercellular material of the desmosome. There are filaments extending from each membrane in the area of the desmosome into the cytoplasm. Desmosomes are common on the aspects of cells of epithelia where like cells are in contact.

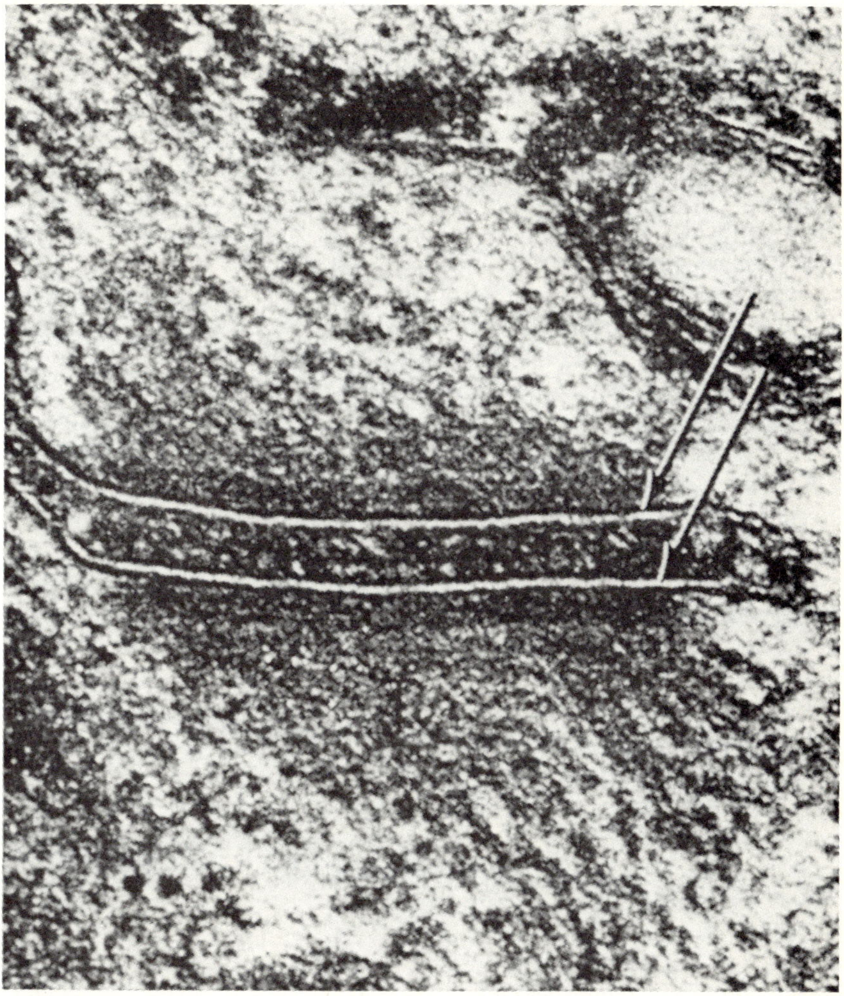

Figure 7-2 Transmission electron microscopic image of a desmosome sectioned perpendicularly to the planes of the associated plasma membranes. The arrows indicate the translucent central laminae of the plasma membranes. The length of the arrows represents 0.1 μm. (Reprinted with permission from *Prog. Biophys. Mol. Biol.,* vol. 26, N. S. McNutt and R. S. Weinstein, Membrane Ultrastructure at Mammalian Intercellular Junctions, Copyright 1973, Pergamon Press, Ltd.)

The morphological correlate of the barrier to transport between extracellular compartments is the *tight junction,* also called the *zonula occludens.* In Section 4-4 a simple form of multicellular system was considered with respect to transport functions of a compartmented system of membranes. The example was an epithelium formed of a monolayer of cells. Such epithelia are arranged in cylinders (as in a renal tubule or intestinal mucosa) or in sacs [as in acinar glands such as exocrine pancreas (Section 5-3)]. These epithelia serve to separate substances by specific secretion or absorption. The major part of this function, if not all of it, is achieved by the plasma membranes of the epithelial cells. Transport between the cells from one side of the epithelium to the other is prevented—at least greatly reduced—by the tight junction.

The tight junction forms a continuous seal around the cells of the epithelium by joining adjacent cells at the ends near the lumen. Examples of a tight junction are at the apical end of the pancreatic acinar cell near the space continuous with the secretory duct system, and at the bases of the microvilli of the intestinal mucosal cell near the lumen. In the tight junction the adjacent plasma membranes are so close that no space is seen between them (see Figure 7-3). In fact, thin-section electron microscopy produces images of cross sections of tight junctions in which the distance across the joined membranes is slightly less than twice the apparent thickness of the plasma membrane of that type of cell. Some biologists describe the tight junction as a fusion of membranes. "Fusion," in this case, has a different meaning from that in Chapter 6.

(In some cases transport across epithelia may involve a path between cells of an epithelium. Such a model has been used to explain the characteristics of H_2O and Na^+ transport in frog skin. The path may be only partly between cells, in which case the substrate does cross the plasma membrane. If such a mechanism exists in any mammalian system it is probably of relatively little quantitative significance.)

The morphological correlate of connections between cytoplasms of adjacent cells is the *gap junction.* At a gap junction the two participating cell membranes are distinct and separated by 2–3 nm (see Figure 7-4). There is some material seen by electron microscopy in this narrow gap that is more dense than background. When a liquid carrying an electron-dense solute is introduced into this system before preparation for electron microscopy, it enters the space between the plasma membranes. The extreme opacity to electrons makes it possible to see that the stain surrounds structures a few nanometers in diameter crossing the gap perpendicularly to the membranes (Figure 7-5). These structures correspond to arrays of projections and complementary pits seen face on in freeze-fracture preparations (see Figure 7-6).

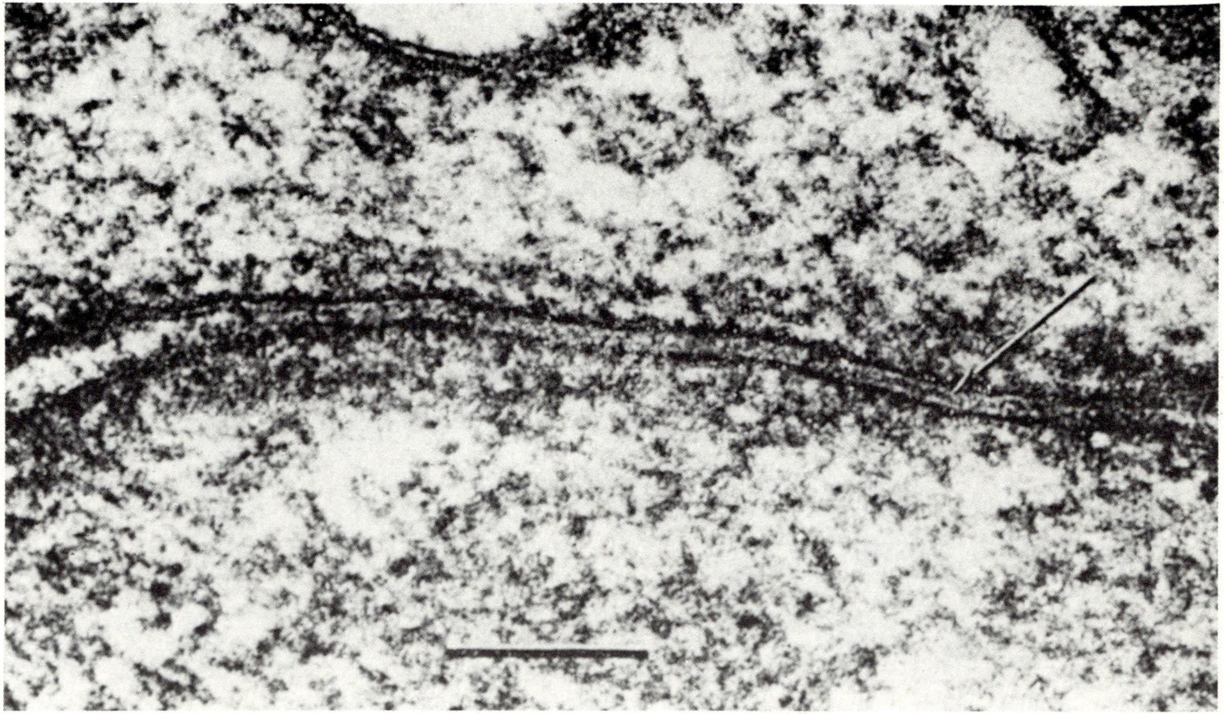

Figure 7-3 Transmission electron microscopic image of a tight junction (arrow) of the epithelium of the proximal tubule of a mouse kidney sectioned perpendicularly to the planes of the adjacent cell membranes. Bar represents 0.1 μm. (Reprinted with permission from *Prog. Biophys. Mol. Biol.*, vol. 26, N. S. McNutt and R. S. Weinstein, Membrane Ultrastructure at Mammalian Intercellular Junctions, Copyright 1973, Pergamon Press, Ltd.)

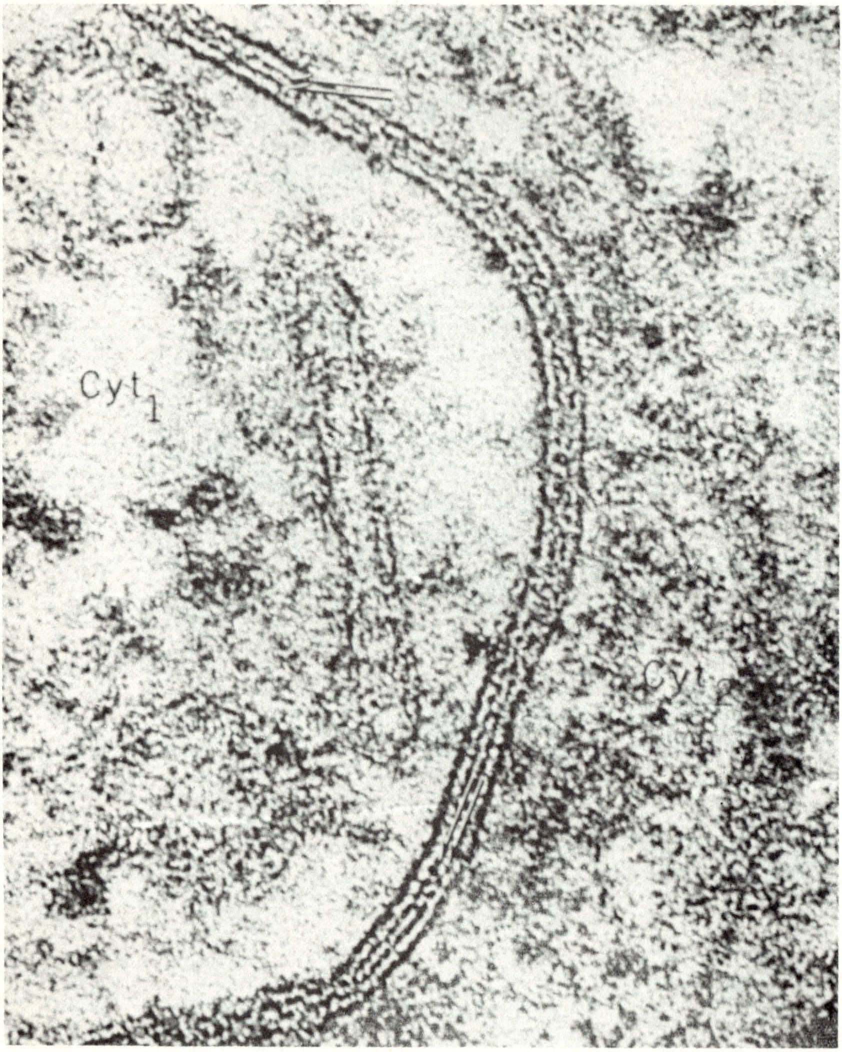

Figure 7-4 Transmission electron microscopic image of a gap junction between two epithelial cells of human cervix sectioned perpendicularly to the planes of the associated plasma membranes. Cyt_1 and Cyt_2 indicate the cytoplasms of the two cells. Arrows indicate the intermembrane separation of approximately 2 nm. (Reprinted with permission from *Prog. Biophys. Mol. Biol.*, vol. 26, N. S. McNutt and R. S. Weinstein, Membrane Ultrastructure at Mammalian Intercellular Junctions, Copyright 1973, Pergamon Press, Ltd.)

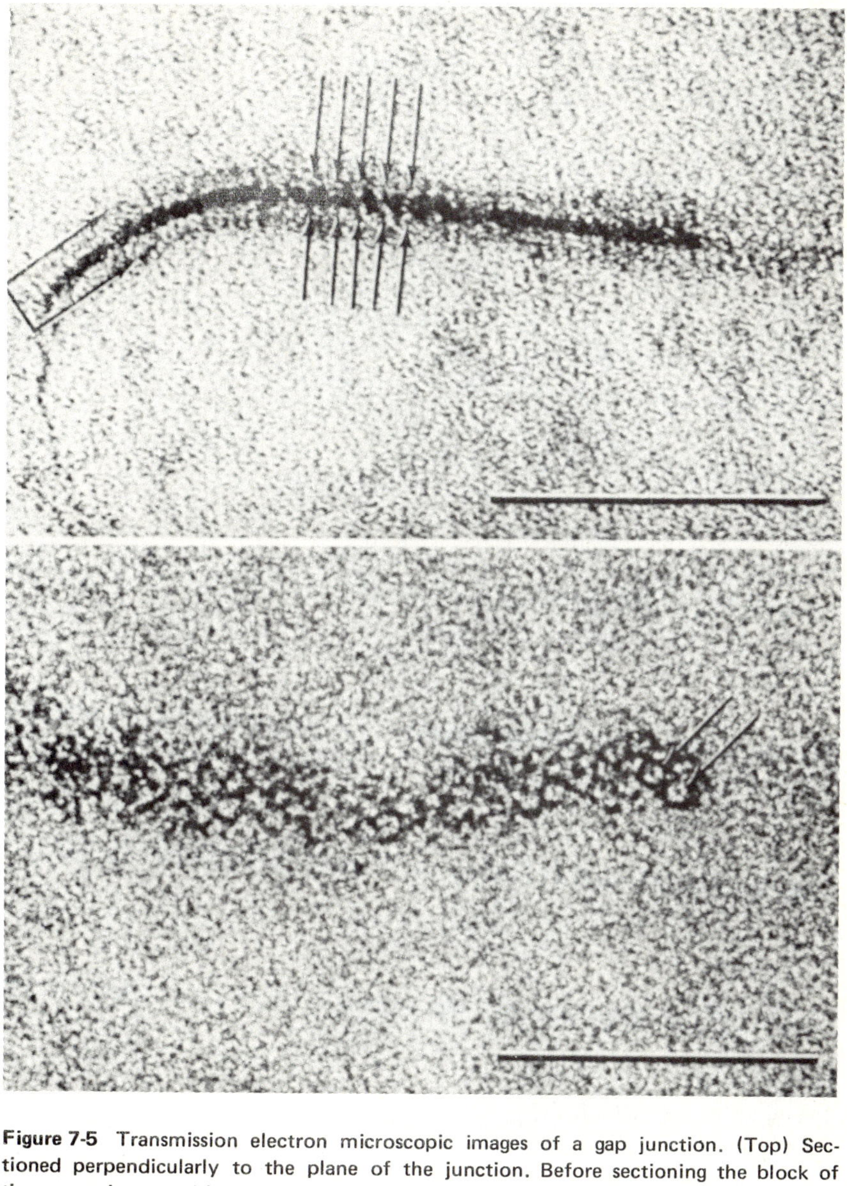

Figure 7-5 Transmission electron microscopic images of a gap junction. (Top) Sectioned perpendicularly to the plane of the junction. Before sectioning the block of tissue was immersed in an aqueous solution containing colloidal lanthanum hydroxide. The electron-opaque La flows through the junction, negatively staining associated structures. (Bottom) Sectioned in the plane of the junction. La reveals structures across the intercellular space. See Figures 7-6 and 7-7. (Reprinted with permission from ***Prog. Biophys. Mol. Biol.,*** vol. 26, N. S. McNutt and R. S. Weinstein, Membrane Ultrastructure at Mammalian Intercellular Junctions, Copyright 1973, Pergamon Press, Ltd.)

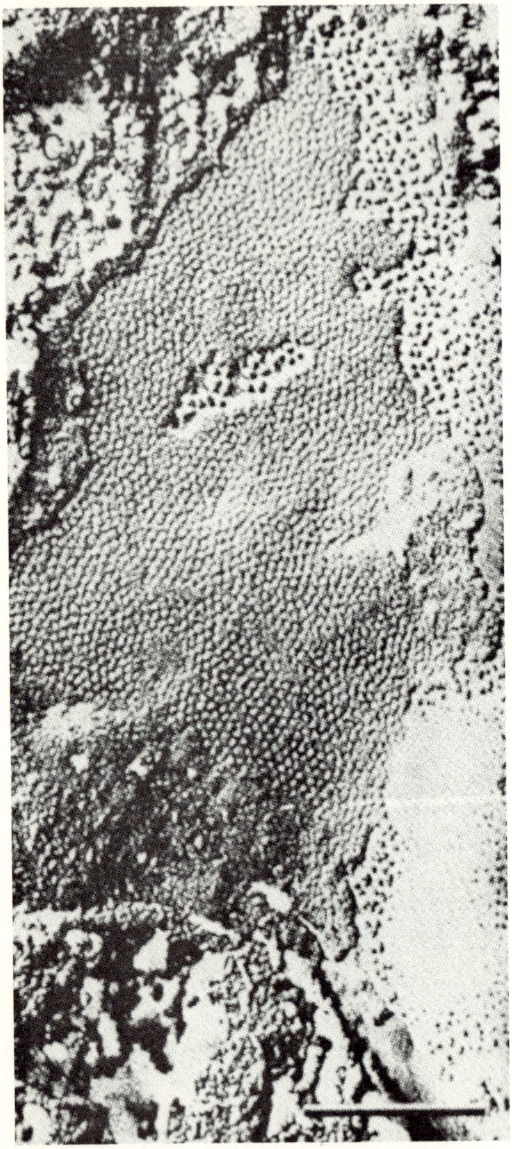

Figure 7-6 Freeze-fracture replica electron microscopic image of a gap junction in mouse heart muscle. The closely packed pits to the left probably correspond with the projecting particles on the right. The heterogeneous area at upper left and lower left represent cytoplasm. Bar represents 0.1 μm. (Reprinted with permission from *Prog. Biophys. Mol. Biol.*, vol. 26, N. S. McNutt and R. S. Weinstein, Membrane Ultrastructure at Mammalian Intercellular Junctions, Copyright 1973, Pergamon Press, Ltd.)

The key to the physiological role of the gap junction is revealed in two experimental explorations of relations between cells sharing gap junctions. As seen in Chapter 3, there is a high electric resistance between an electrode inside a cell and one outside the cell in a conducting physiological solution. This high resistance is attributable to the plasma membrane. A similar high resistance exists between two adjacent intact cells when there are no gap junctions between them. However, when gap junctions exist, the resistance is reduced so as to be essentially indistinguishable from that of the medium or the cytoplasm. The presence of gap junctions is, therefore, seen to be associated with a high electric conductivity between the cytoplasms of the associated cells; the gap junctions permit free passage of ions (see Figure 7-7).

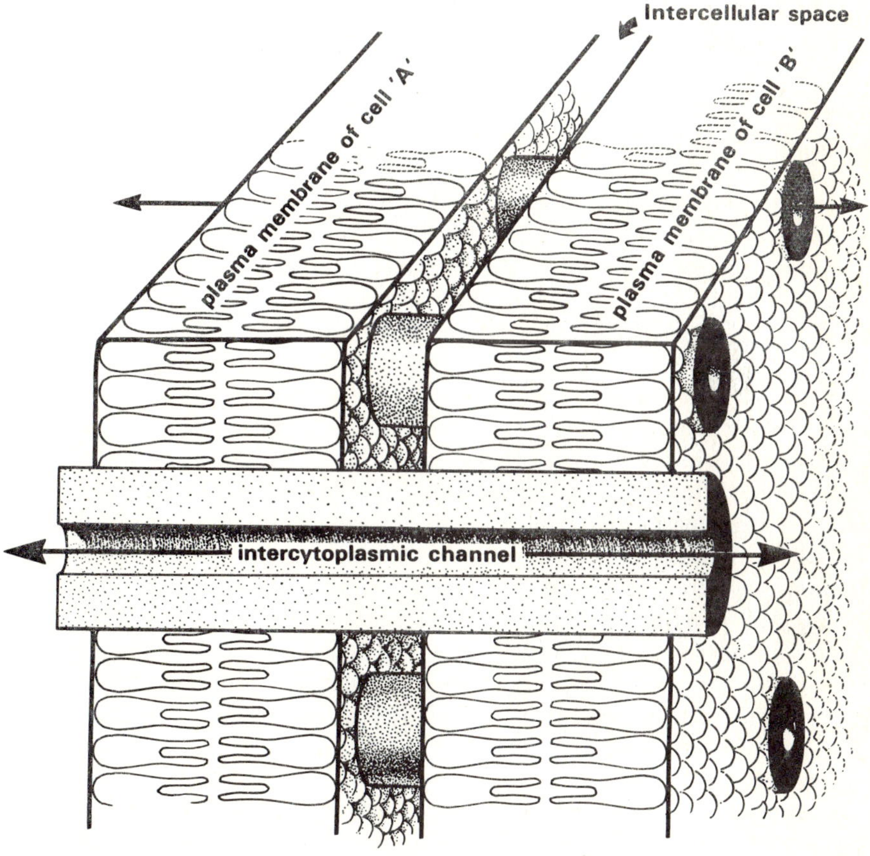

Figure 7-7 Schematic representation of the structure of a gap junction. The plasma membranes of cells A and B are 2–3 nm apart.

The other experimental demonstration is similar but goes further. It is possible to inject material into epithelial cells sharing gap junctions with adjacent cells. One can inject identifiable molecules of various sizes (fluorescein, for example, detectable by its fluorescence). For molecules of a molecular weight up to the order 10^3 daltons, passage into cells adjacent to the injected cell is seen where gap junctions are present. These experiments seem to show that cells connected by gap junctions form a syncytium—that is, they form a multinuclear cytoplasmic compartment connected for all but the largest molecules and ions. This situation must serve to integrate the states and activities of the component cells of entire epithelia or of domains of many cells within an epithelium.

Gap junctions are not necessarily static features of a cell type. Functionally they close with respect to intercytoplasmic transport in response to particular kinds of injury of one of the participating cells. Electric resistance rises to the closed-membrane level; Ca^{2+} and small organic molecules are unable to pass. When cells that normally form gap junctions are allowed to make contact under physiological conditions, gap junctions form. This sort of behavior certainly seems consistent with greater survival value. The mechanism of such response to circumstances is not known. This adaptiveness, morphological evidence, and precedents from other membrane-mediated phenomena indicate that these junctions probably form by a coordinated rearrangement of the integral proteins of both involved plasma membranes in concert with cytoplasmic events.

7-5 SUMMARY

Intercellular relations include chemical communication between cells at some distance and interactions at regions of contact. Hormones and neurotransmitter substances are prominent examples of interaction by means of substances dissolved in extracellular liquid. The peptide hormones first influence a cell by binding to specific receptors on the plasma membrane. This reaction may alter some property of the plasma membrane such as permeability to a particular substrate. The membrane, probably through a specific integral protein, conveys the signal to the intracellular responsive components in some cases by activating adenyl cyclase, which is a membrane-bound lipoprotein. The number of hormone receptors and their affinity for hormone are subject to alteration in response to extracellular hormone concentration.

Reactions to cell contact include agglutination, which may be nonspecific or quite specific with respect to receptors and inducers, and formation of special intercellular junctions. These junctions include desmosomes that seem to fix cell to cell, tight junctions that fix cell to cell in epithelia and also seal intercellular pathways, and gap junctions that connect cytoplasms of adjacent cells with channels permeable to small ions and molecules. Gap junctions respond to cell disruption by closure of these channels.

Surface components, such as glycoproteins and glycosyltransferases, vary with differentiation and morphogenesis during embryogenesis. Cell recognition, differentiation, intercellular specific adhesion, and morphogenesis involve temporal and spatial variations in the chemical composition and organization of plasma membranes.

Annotated Bibliography

The references included here are only books and review papers—not reports of original investigations—on topics that would be considered rather broad in the context of modern specialization.

The first major section consists of two lists, one of books and one of review articles, which are given in alphabetical order by author or editor and which are numbered for easy cross-reference. The second major section includes topic headings under which are numbers of the appropriate book or paper, comments on the references and cross-references to related chapters or sections in this book.

There are now many books on biomembranes, but there are very few designed as texts that include all aspects. Most are collections of papers on very special topics. One good, concise book on biomembranes is that by P. J. Quinn, "The Molecular Biology of Cell Membranes," University Park Press, Baltimore, 1976. A good, up-to-date text in cell biology will be helpful for following up some of the topics or questions raised in this book. Such a book is one by R. D. Dyson, "Cell Biology, a Molecular Approach," 2d ed. Allyn & Bacon, Boston, 1978.

REFERENCES

Books

B1 Brinkley, B. R. and K. R. Porter (eds.): "International Cell Biology," Rockefeller University Press, 1977.

B2 Bronner, F. and A. Kleinzeller (eds.): "Cell Surface Glycoproteins: Structure, Biosynthesis and Biological Functions," Academic, New York, 1978 (Vol. 11 of *Curr. Top. Memb. Transp.*).

B3 Christensen, H. N.: "Biological Transport," 2d ed., W. A. Benjamin, Don Mills, Ontario, 1975.

B4 Fermi, E.: "Thermodynamics," Dover, New York, 1956.

B5 Fox, C. F. (ed.): "Biochemistry of Cell Walls and Membranes," Butterworths, London, 1975.

B6 Katchalsky, A. and P. F. Curran: "Nonequilibrium Thermodynamics in Biophysics," Harvard University Press, Cambridge, 1965.

B7 Katz, B.: "Nerve, Muscle and Synapse," McGraw-Hill, New York, 1966.

B8 Ling, G. N.: "A Physical Theory of the Living State: The Association-Induction Hypothesis," Blaisdell, New York, 1962.

B9 Poste, G. and G. L. Nicolson: "The Cell Surface in Animal Embryogenesis and Development," North-Holland, New York, 1976 (vol. 1 of "Cell Surface Reviews").

B10 Poste, G. and G. L. Nicolson (eds.): "Dynamic Aspects of Cell Surface Organization," North-Holland, New York, 1977 (vol. 3 of "Cell Surface Reviews").

B11 Poste, G. and G. L. Nicolson (eds.): "The Synthesis, Assembly and Turnover of Cell Surface Components," North-Holland, New York, 1977 (vol. 4 of "Cell Surface Reviews").

B12 Prigogine, I.: "Introduction to Thermodynamics of Irreversible Processes," Wiley-Interscience, New York, 1961.

B13 Tanford, C.: "The Hydrophobic Effect: Formation of Micelles and Biological Membranes," Wiley, New York, 1973.

B14 Weissmann, G. and R. Claiborne (eds.): "Cell Membranes: Biochemistry, Cell Biology and Pathology," Hospital Practice, New York, 1975.

Review Papers

R1 Bradshaw, R. A. and W. A. Frazier: Hormone receptors as regulators of hormone action, *Curr. Top. Cell. Regul.* 12: 1, 1977.

R2 Case, R. M.: Synthesis, intracellular transport and discharge of exportable proteins in the pancreatic acinar cell and other cells, *Biol. Rev.* 53: 211, 1978.

R3 Coleman, R.: Membrane-bound enzymes and membrane ultrastructure, *Biochim. Biophys. Acta* 300: 1, 1973.

R4 DePierre, J. W. and L. Ernster: Enzyme topology of intracellular membranes, *Annu. Rev. Biochem.* 46: 201, 1977.

R5 Marchase, R. B., K. Visbeck, and S. Roth: Intercellular adhesive specificity, *Biochim. Biophys. Acta* 457: 385, 1976.

R6 McNutt, N. S. and R. S. Weinstein: Membrane ultrastructure at mammalian intercellular junctions, *Prog. Biophys. Mol. Biol.* 26: 47, 1973.

R7 Mitchell, P.: Vectorial chemistry and the molecular mechanics of chemiosmotic coupling: Power transmission by proticity, *Biochem. Soc. Trans.* 4: 399, 1976.

R8 Neville, D. M., Jr., and T. A. Chang: Receptor-mediated protein transport into cells. Entry mechanisms for toxins, hormones, antibodies, viruses, lysosomal hydrolases, asialoglycoproteins and carrier proteins, in F. Bronner, and A. Kleinzeller (eds.), "Membrane Properties: Mechanical Aspects, Receptors, Energetics and Calcium-Dependence of Transport, Academic, New York, 1978 (vol. 10 of "Curr. Top. Membr. Transp.").

R9 Phillips, M. C.: The physical state of phospholipids and cholesterol in monolayers, bilayers, and membranes, *Prog. Surf. Membr. Sci.* 5: 139, 1972.

R10 Poste, G. and A. C. Allison: Membrane fusion, *Biochim. Biophys. Acta* 300: 421, 1973.

R11 Silverstein, S. C., R. M. Steinman, and Z. A. Cohn: Endocytosis, *Annu. Rev. Biochem.* 46: 699, 1977.

TOPICS COVERED

History (Section 1-1)

A short account of recent developments specifically related to biomembranes is given in B2 by A. Rothstein (pp. 1–13).

Thermodynamics

There are many books on thermodynamics written from a variety of viewpoints, and physical chemistry texts include thermodynamics as a major component. One needs to browse and choose. B4 is a short, clear, classical treatment written over 40 years ago by a great physicist who was also a great teacher. One of the more manageable expositions of the thermodynamics of irreversible processes is B12, which was written by a pioneer developer of the subject. B6 also introduces the subject and relates it to biologically significant phenomena.

Lipids, Micelles, and Amphiphilic Monolayers (Chapter 2)

R9 is an excellent treatment of the physicochemical behavior of the lipids prominent in biomembranes. B13 concisely presents the thermodynamics of micelle formation. Chapters 12–19 are of particular interest to biologists and can be read without having mastered the physical chemistry of preceding chapters.

Molecular Organization and Component Mobility (Chapter 2)

B14 is a well-illustrated collection of brief reviews of special topics. The authors are leading authorities. The first six papers constitute a clear introduction to this topic: Danielli on the bilayer hypothesis, Chapman on lipid dynamics, Bangham on real artificial models of biomembranes, Singer on molecular organization, Marchesi on the relations of an exemplary glycoprotein to the membrane, and Roseman on the carbohydrates of the plasma membrane. A more general and more rigorous discussion of the orientation of integral proteins is given by Marchesi in B5. Component mobility is emphasized in a contribution to B10 by Nicolson, Poste, and Ji.

Turnover of Membrane Constituents (Chapter 5)

R. T. Schimke has a paper in B5 on the turnover of membrane proteins in which a section on the turnover of lipid constituents is included. The general remarks and comments on methodology are particularly useful. In a paper in B11, J. Tweto and D. Doyle discuss the turnover of proteins of the eucaryotic cell surface with emphasis on the kinds of evidence and problems of interpretation.

Membrane Glycoproteins (Sections 2-3 and 2-4)

B2, a 1978 collection of reviews, is a rich source on glycoproteins and brings out clearly the wide variety of biologic functions of membrane glycoproteins. Topics in this volume include structure and biosynthesis (Sturgess, Moscarello, and Schachter), analytic techniques (Juliano), and erythrocyte glycoproteins (Tanner). In B11 G. M. W. Cook describes the biosynthesis of membrane glycoproteins.

Membrane Enzymes (Sections 4-2 and 4-3)

R3 is somewhat dated, but the organization and content of this review provides much insight into the special features of enzymes as components of biomembranes. In B2 J. R. Riordan and G. G. Forstner present a detailed review of glycoprotein membrane enzymes. In B10 H. K. Kimelberg discusses the relation between membrane fluidity and the activity of membrane-bound enzymes. R4 is a detailed review of the topology of enzymes in the various intracellular membranes.

Membrane Biogenesis (Chapter 5)

In B14 B. F. Trump discusses the possible interrelations among membranes that are central in hypotheses of biogenesis, and J. D. Jamieson discusses secretion in a paper that is relevant to biogenesis. A leader in research in this area, D. J. Morré, reviews the problem in B11. Also in B11, the papers by G. M. W. Cook and by J. Tweto and D. Doyle are very useful, and Holtzman, Schacher, Evans, and Teichberg deal with biogenesis when considering various secretory vesicles.

Transmembrane Transport (Chapter 3)

Most physiology and cell biology books have cursory expositions on this topic; they vary widely in quality. Transport from the viewpoint of thermodynamics of irreversible processes [the Kedem-Katchalsky approach orginally published in 1958 (*Biochim. Biophys. Acta* 27: 229)], along with the underlying physics, is treated in B6. This is a short but tightly packed book. B3 includes both thermodynamic and kinetic approaches. Being a larger book, it gives more emphasis to experimental methods and biologic examples.

Sorption Theory (Section 3-3)

This controversial topic is not widely represented in textbooks. B8, a large book devoted to this subject, was published in 1962. It was written by a major advocate and developer of the hypothesis, G. N. Ling, who is also a productive experimenter. A 1977 paper by Ling (*Mol. Cell. Biochem.* 15: 159) is a summary of this theory, and it gives more recent references plus a new offshoot of the hypothesis.

Excitable Tissues and Electrophysiology (Section 3-2)

B7 is a short, clear book about nerve and muscle excitability and synaptic transmission. B. Katz is an authority on the subject and has written a fine exposition that can be read by a novice for simple pleasure and acquaintance or by a specialist for critical analysis of evidence. Many physiology books have useful complete treatments of this topic.

Mitochondria (Section 4-3)

There are weighty tomes solely about this organelle. R7 is a review paper by P. Mitchell who has been a moving force in theoretical and experimental understanding of the role of the biomembrane in photosynthetic and oxidative generation of energy-rich compounds. Mitchell is generally recognized as the originator of the chemiosmotic hypothesis. This paper predates some important experimental results, but it contains the essentials. R4 includes a detailed discussion of mitochondrial membrane enzymes. In B14 E. Racker gives a short account of the field as of 1975.

Cell Fusion (Section 6-3)

R10 is a useful review, although it is too old to include some significant information. "Membrane Fusion" [G. Poste and G. L. Nicolson (eds.), North-Holland, New York] is to appear soon. It will contain papers on general features of membrane fusion as well as cases peculiar to various biologic phenomena. J. A. Lucy gives a brief discussion of fusion in B14.

Exocytosis (Section 6-4)

R2 is an excellent review of protein secretion with emphasis on the pancreatic acinar cell. In it Case focuses on the separable steps of secretion including the fusion of secretory vesicle with plasma membrane. Jamieson in B14 gives a brief discussion of this topic, and B11 includes a pertinent review by Holtzman, Schacher, Evans, and Teichberg.

Endocytosis (Section 6-4)

R11 is a detailed treatment of endocytosis. R8 includes some important aspects of receptors and carriers in pinocytosis and some discussion of general models. In B11 Elspach has a discussion of phagocytosis in which the role of the plasma membrane is emphasized. Korn, in B5, presents a systematic view of endocytosis and reviews what was then known for mammalian cells and amoebas.

Mitosis (Section 6-5)

A discussion of the role of the plasma membrane in cytokinesis and control of mitosis by A. B. Pardee and E. Rozengurt appears in B5. J. G. Bluemink and S. W. de Laat treat the variation in plasma membrane related to the cell cycle, including cell division, in B11. A short paper by J. M. Arnold in B9 focuses on membrane-related problems in cytokinesis.

Hormone-Membrane Interactions (Section 7-1)

This is currently the focus of much experimental work and considerable controversy. Evidence is sometimes conflicting and often interpretable in contradictory ways. There will be no up-to-date review for some years, but R1 is very nicely done, gives the major evidence and arguments, and poses the questions well. In B14 P. Cuatrecasas' short paper presents the problem clearly and shows where certain experimental methods fit in; in the light of more recent information the conclusions may be overly simple. Again in B14, N. D. Goldberg discusses cyclic nucleotides and cyclases in their putative roles; the mechanism he proposes is clearly not as universal as suggested, and newer evidence clouds the picture. The article, however, places the hormone-receptor interaction into the context of the total cellular response.

Intercellular Adhesion (Section 7-2)

R5 is a wide-ranging review, giving a survey of biologic events in which adhesion plays a central role and including a discussion of methodology. In B2 L. A. Culp contributes a good discussion of cell adhesion to non-cellular substrates and to other cells. G. Gerisch in B1 has a short paper on slime mold membranes involved in cell adhesion. His discussion is very much in the spirit of the following topic.

Cell Surfaces and Development (Section 7-3)

Gerisch in B1 was noted above. B9 is a collection of reviews devoted to fertilization, implantation, and development. Of particular interest with respect to membranes are the papers by Edidin on surface antigens, Sheridan on cell coupling and communication, and Maslow on specific adhesion among embryonic cells. Recent books on embryology often include material on cell surface chemistry.

Intercellular Junctions (Section 7-4)

In B14 G. D. Pappas gives a very brief description of the three types of junctional complexes. (There is a mistake in the schematic of the gap junction; the evidence seems to preclude bilayer enclosure of the transgap channel structures.) Also in B14 W. R. Loewenstein has a very short, lucid, and well-illustrated discussion of the gap junction, which emphasizes its permeability role. Sheridan in B9 discusses functional and structural features of gap junctions and experimental methods, as well as the variations in embryogenesis with respect to gap junctions. B1 includes a section entitled "Cell-to-Cell Communications," in which Gilula presents a brief discription of gap junctions that includes some very fine electron microscopic pictures; Loewenstein describes, more technically than in B14, his intercellular conductance experiments; and Sotelo discusses electric (gap junction) and chemical (synaptic) communication in the central nervous system. A very detailed description of all three classes of junctions is given in R6, which includes technical considerations of methodology and some good electron microscopic pictures.

Lymphocytes

The immunology literature on lymphocytes is abundant, and new developments have been rapid. The role of the cell surface and cell membrane components is emphasized in B10 in reviews by Sundqvist, Andersson and Melchers, and de Petris. A very brief glimpse of lymphocyte membranes is provided by J. W. Uhr in B14.

Index